T0215802

Design, Representations, and Processing for Additive Manufacturing

Synthesis Lectures on Visual Computing: Computer Graphics, Animation, Computational Photography, and Imaging

Editor

Brian A. Barsky, *University of California, Berkeley*

This series presents lectures on research and development in visual computing for an audience of professional developers, researchers, and advanced students. Topics of interest include computational photography, animation, visualization, special effects, game design, image techniques, computational geometry, modeling, rendering, and others of interest to the visual computing system developer or researcher.

Mathematical Basics of Motion and Deformation in Computer Graphics, Second Edition
Ken Anjyo and Hiroyuki Ochiai
2017

Digital Heritage Reconstruction Using Super-resolution and Inpainting
Milind G. Padalkar, Manjunath V. Joshi, and Nilay Ł. Khatri
2016

Geometric Continuity of Curves and Surfaces
Przemyslaw Kiciak
2016

Heterogeneous Spatial Data: Fusion, Modeling, and Analysis for GIS Applications
Giuseppe Patanè and Michela Spagnuolo
2016

Geometric and Discrete Path Planning for Interactive Virtual Worlds
Marcelo Kallmann and Mubbasir Kapadia
2016

An Introduction to Verification of Visualization Techniques
Tiago Etiene, Robert M. Kirby, and Cláudio T. Silva
2015

Virtual Crowds: Steps Toward Behavioral Realism
Mubbasir Kapadia, Nuria Pelechano, Jan Allbeck, and Norm Badler
2015

Finite Element Method Simulation of 3D Deformable Solids
Eftychios Sifakis and Jernej Barbič
2015

Efficient Quadrature Rules for Illumination Integrals: From Quasi Monte Carlo to Bayesian Monte Carlo
Ricardo Marques, Christian Bouville, Luís Paulo Santos, and Kadi Bouatouch
2015

Numerical Methods for Linear Complementarity Problems in Physics-Based Animation
Sarah Niebe and Kenny Erleben
2015

Mathematical Basics of Motion and Deformation in Computer Graphics
Ken Anjyo and Hiroyuki Ochiai
2014

© Springer Nature Switzerland AG 2022

Reprint of original edition © Morgan & Claypool 2018

All rights reserved. No part of this publication may be reproduced, stored in a retrieval system, or transmitted in any form or by any means—electronic, mechanical, photocopy, recording, or any other except for brief quotations in printed reviews, without the prior permission of the publisher.

Design, Representations, and Processing for Additive Manufacturing

Marco Attene, Marco Livesu, Sylvain Lefebvre, Thomas Funkhouser, Szymon Rusinkiewicz, Stefano Ellero, Jonàs Martínez, and Amit Haim Bermano

ISBN: 978-3-031-01468-0   paperback
ISBN: 978-3-031-02596-9   ebook
ISBN: 978-3-031-00347-9   hardcover

DOI: 10.1007/978-3-031-02596-9

A Publication in the Springer series
Synthesis Lectures on Visual Computing: Computer Graphics, Animation,
Computational Photography, and Imaging

Lecture #31
Series Editor: Brian A. Barsky, *University of California, Berkeley*
Series ISSN
Print 2469-4215   Electronic 2469-4223

Design, Representations, and Processing for Additive Manufacturing

Marco Attene and Marco Livesu
CNR–IMATI Genova

Sylvain Lefebvre and Jonàs Martínez
INRIA

Thomas Funkhouser, Szymon Rusinkiewicz, and Amit Haim Bermano
Princeton University

Stefano Ellero
STAM S.r.l

SYNTHESIS LECTURES ON VISUAL COMPUTING: COMPUTER GRAPHICS, ANIMATION, COMPUTATIONAL PHOTOGRAPHY, AND IMAGING #31

ABSTRACT

The wide diffusion of 3D printing technologies continuously calls for effective solutions for designing and fabricating objects of increasing complexity. The so-called "computational fabrication" pipeline comprises all the steps necessary to turn a design idea into a physical object, and this book describes the most recent advancements in the two fundamental phases along this pipeline: design and process planning. We examine recent systems in the computer graphics community that allow us to take a design idea from conception to a digital model, and classify algorithms that are necessary to turn such a digital model into an appropriate sequence of machining instructions.

KEYWORDS

geometric modeling, geometry processing, 3D printing

Contents

Acknowledgments

Marco Attene, Marco Livesu, and Stefano Ellero's work for this book was partly funded by the European Union's Horizon 2020 research and innovation programme under grant agreement No. 680448 (CAxMan). Thanks are due to all the participants in CAxMan, to the members of the Shapes and Semantics Modeling Group at IMATI-CNR, and to colleagues at STAM S.r.l. for helpful discussions. Sylvain Lefebvre and Jonàs Martínez's work was partly supported by ERC grant ShapeForge (StG-2012-307877) and ANR MuFFin (ANR-17-CE10-0002). All the authors thank the reviewers for their helpful comments and suggestions.

The work of Amit H. Bermano, Thomas Funkhouser and Szymon Rusinkiewicz for this book was partially funded by the DARPA agreement no. HR0011-15-2-0040. These authors would also like to thank Justin Solomon, Bill Regli, and Jan Vandenbrande for many helpful conversations, as well as Sema Berkiten for helping in summarizing some of the papers.

May 2018

CHAPTER 1

Introduction

In its standard definition (ISO/ASTM 52900), Additive Manufacturing (AM) is *the process of joining materials to make parts from 3D model data, usually layer upon layer, as opposed to subtractive manufacturing and formative manufacturing methodologies*. Thanks to its flexibility, Additive Manufacturing is rapidly being adopted in industrial practice for product realization, either as a replacement or in conjunction with more traditional subtractive technologies.

1.1 COMPUTATIONAL FABRICATION

Any modern industrial product has a typical lifecycle constituted of the four successive phases of *conception*, *design*, *realization*, and *service*: once a product concept is designed by a CAD expert, the resulting 3D model must be realized through proper fabrication tools, and then put into service for actual use.

When compared with traditional technologies, AM provides enormous freedom to the designer that may specify virtually any sort of geometry, without struggling too much with its realizability. On the contrary, the real problem today is the lack of tools to actually unlock this freedom. This situation triggered the emergence of a new design concept, where functional performance, manufacturability, reliability and cost are all optimized based on the capabilities of AM technologies. In some cases, the traditional scenario where a designer creates the part geometry is completely reversed and, instead of defining the geometry, the designer is called to specify physical requirements that, in turn, drive an automatic synthesis of the final shape (e.g., in the so-called "topology optimization").

In any case, the design model must be analyzed, and the process necessary for its physical realization must eventually be determined. The sequence of operations required to move from design to realization is known as Process Planning (PP). For AM technologies, a typical PP includes a tessellation step, the definition of a building direction, the calculation of parallel slices, and the conversion of the slices into commands for the printer. In some cases additional operations can be necessary, such as the calculation of support structures or the partitioning of large objects that do not fit the printing chamber.

Creating tools to support this Computational Fabrication pipeline requires a deep understanding of 3D geometry, and that is why the Computer Graphics community is so intensively contributing to this important area. Expectedly, this research is extremely active in Manufacturing and Material Science areas too, where the focus is mostly on physical properties of the process. In this book, we examine and discuss all these contributions while focusing on Design and Process Planning, and outline challenging problems for which further research is still necessary.

1.2   CONTENT AND OBJECTIVES

Our main objective is to provide the reader with a comprehensive overview of the state of the art in Design and Process Planning for Additive Manufacturing. The main focus is on the modeling, processing, and analysis of the geometry, but other characteristics are taken into account along the path (e.g., mechanical and thermal behavior, appearance, response to stresses, electrical conductivity, etc.). Although we try not to focus on any specific printing technology, when process planning is concerned most recent literature deals with plastic-based AM and, to a smaller extent, with metal sintering. Nonetheless, the AM field is very wide and includes applications ranging from food printing to AM-aided construction; these topics, however, are out of scope for this book. For a more detailed description of these technologies, we point the reader to the survey by Julien Gardan [Gardan, 2016].

Chapter 2 starts by introducing the most diffused 3D printing technologies and by classifying them into broad families, each characterized by a common set of needs, possibilities, and limitations. Then, the computational fabrication pipeline is analyzed, and the possible objectives that a printing process must reach are summarized, along with existing approaches to find the best tradeoff, also depending of the specific technology at hand.

Then, Chapter 3 reports the various design concepts invented so far to best exploit the possibilities offered by AM. The report is based on design objectives that range from the realization of a specific appearance, through physical robustness and elasticity, up to high-level goals such as functionality and utility.

Chapter 4 enters the core of the process planning phase by analyzing the details of each step along the conversion of a design model to a sequence of machine instructions. We deal with the compatibility of a design model with a given printing technology, and report on algorithms that cover all the main steps such as orientation, slicing, support structures, and infill generation. Also, the impact of these algorithms on the printing objectives is analyzed.

Finally, Chapter 5 summarizes the problems that are to be considered still open and particularly challenging.

This book is the result of a joint effort to extend, unify, and update two recent eurographics state-of-the-art reports that covered Design [Bermano et al., 2017] and Process Planning [Livesu et al., 2017] up to early 2017.

CHAPTER 2

Practices and Considerations for Additive Manufacturing

2.1   3D PRINTING TECHNOLOGIES

In this section, we briefly describe the main additive manufacturing technologies. Our goal is not to provide an extensive list, but rather to discuss the main properties—and limitations—that impact the process planning.

All the technologies we consider build an object layer after layer. They mainly differ, however, by whether they actually *locally deposit* material or whether they *solidify material* within an otherwise non-solid substance. This has important implications on the process planning and this distinction therefore is the basis of our two main categories: *material deposition* and *layer solidification*.

A second fundamental distinction between these technologies is whether they deposit/solidify material along continuous paths (vector) or whether they rely on a discrete device (raster). This directly drives whether the output of the process planning is a set of continuous paths (vector) or a set of images (raster).

Material Deposition

Material deposition refers to methods that create the next layer by *locally* depositing material on a previously printed layer. This encompasses techniques such as material extrusion (e.g., fused filament deposition [Crump, 1989]), material jetting (e.g., UV-sensitive resin droplets [PolyJet, 1998, Sitthi-Amorn et al., 2015]), and directed energy deposition (e.g., laser cladding [Vilar, 1999]).

Vector or raster.   Filament-fused deposition is a vector approach as it deposits continuously along paths. The motion of the extruder is achieved through either a three axis gantry, or a delta robot configuration. Processes relying on resin droplets are usually discrete, similarly to inkjet printers. The print head is often attached to a two-axis gantry, with the build plate moving up or down along the layering direction.

Properties.   The key advantages of material deposition are the ability to combine multiple materials, a printing time that mostly depends on the part volume, and the ability to fully enclose voids. A major inconvenient however is the strong requirement for support structures, since material can only be deposited on top of an already existing layer. The process planning therefore has to automatically generate disposable support structures, which we discuss in Section 4.4.1.

Finally, some of these technologies are able to print *out-of-plane*, for instance generating a continuous spiralling path from bottom to top (e.g., *spiralize* feature of the *Cura* open source slicer) or even wire-frame structures [Mueller et al., 2014]. We discuss this in more details in Section 4.5.2

Layer Solidification

Layer solidification refers to all the processes that build the object by solidifying the top (or bottom) surface of a non-solid material (powder, liquid), typically within a tank. This starts by lowering the tank, adding a full layer of non-solid material, and then using a process that solidifies the material in specific places. This encompasses technologies such as vat photo-polymerization (e.g., stereolithography or SLA), powder bed fusion (e.g., selective laser sintering or SLS), binder jetting (e.g., plaster powder binding [Cima et al., 1994]), and sheet lamination (e.g., paper layering—cutting [Mcor, 2005]).

Vector or raster.   SLA processes have both variants, relying either on a laser beam (vector) or on a projected image using a DLP projector (raster). SLS processes are typically driving a laser beam through continuous motions, following contour paths. Both for SLS and SLA, beam motions are obtained using mirrors and galvanometer mechanisms—forming a so-called laser scanner—providing fast and precise movements.

Properties.   A major advantage of layer solidification is the reduced need for support structures on complex geometries, enabling a much wider range of parts to print without any support. Note that supports may still be necessary to stabilize the part (see Section 4.4.1) as it may be able to move within the non-solidified material (in particular with liquid resins, but also in powder depending on part weight). Another need for support arise from heat dissipation issues, in particular with metal powder melting.

A drawback of within-layer solidification is that it is more challenging to mix different materials. This is achieved on some technologies by locally depositing additives, such as pigmented inks on powder-binding 3D printers such as Cima et al. [1994] or the HP Jet Fusion 3200, or by masking techniques using different resin tanks on SLA printers [Zhou et al., 2011]. Another drawback is the necessity for non-solidified materials to exit cavities: this prevents the formation of fully closed empty voids within the part.

Layer solidification presents an interesting tradeoff regarding printing times. Each layer starts by a full-tank layer filling (powder) or resin sweep (SLA) which usually takes the same, constant time. This implies that printing time is much more impacted by the height (number of layers) of the object than by the solidified volume. As a consequence, printing a single small object is generally time consuming, while printing objects in batches can lead to significantly reduced print time per-parts, as the constant per-layer time is amortized. We discuss printing in batches in Section 4.2.3.

Table 2.1: Relation between broad AM technological solutions and supported features

	Vector/ Raster	Multiple Materials	Support Structures	Cavities	Build Time
Material Deposition (e.g., fused filament, resin droplets)	Both	Supported	Overhangs	Supported*	Depends on part volume
Material solidification (e.g., liquid resin, various powders)	Both	Supported (technology dependent)	Stability/ Heat Dissipation	Not Supported	Depends on number of layers

*: assuming the boundary of the internal cavities can be printed without support structures.

2.2   THE COMPUTATIONAL FABRICATION PIPELINE

2.2.1   OVERVIEW

The Computational Fabrication Pipeline is the process that allows users to transform a concept into a design and finally into a physical part. This sub-section briefly outlines the conventional Computational Fabrication Pipeline and highlights the main differences introduced by Additive Manufacturing. Depending on the application that the part targets (e.g., a decorative object or a mechanical component) and the technologies involved (e.g., a CNC milling machine or a 3D printer), the Pipeline may skip some of the steps identified in the following.

Similarly to the conventional subtractive manufacturing case, the Computational Fabrication Pipeline for additive manufacturing (Figure 2.1) starts from design specifications. These are mainly meant to define the geometry of the part to be printed. Further to that, additional requirements are specified, such as: the acceptable dimensional and geometrical tolerances of features (e.g., the diameter of a hole or the parallelism between two axes), the need for surface finishing (i.e., which surfaces need to be smooth or polished, typically to mate different parts), the materials to be used, etc. Industrial practice in this phase is still dominated by 2D technical drawings, where all this information is indicated according to standardized rules. However, software tools (e.g., modelers from Autodesk and Dassault) and standard format specifications (e.g., STEP ISO 10303) exist to produce and represent accurate designs directly in 3D: in this latter case the so-produced models are known as Computer-Aided Design (CAD) models. Either 2D technical drawings or 3D CAD models must normally be cast to a corresponding Computer-Aided Manufacturing (CAM) representation to undergo the fabrication process, although some integrated CAD/CAM systems are emerging to make this transition as transparent as possible.

Thus, in a standard computational fabrication pipeline, moving from the CAD to the CAM world represents the switch from the design to the process planning phase. Based on the CAM model, the final goal of Process Planning is to determine the machining instructions that the fabrication tool must execute to build the part. For example, if a hole has to be drilled, the instructions will include information about the drill bit (i.e., the tool) with the required diameter, the coordinates of the hole axis, the height of the start surface, the hole depth, etc.

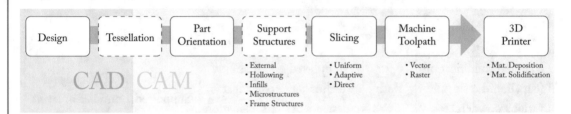

Figure 2.1: The computational fabrication pipeline for Additive Manufacturing at a glance. Boxes with dashed boundaries are to be considered optional [Livesu et al., 2017]. Used with permission.

Due to the layer-by-layer nature of Additive Manufacturing, it is important to select an appropriate building direction (i.e., the direction along which the layers will be built) and to slice the model accordingly, i.e., to decompose the 3D geometry in a set of 2D layers. Each of the slices must then be converted to a proper *toolpath*, that is, to a sequence of movements that the building tool must follow to fabricate the slice: these movements track the outer boundary of the slice, but also its inner parts and possible support structures. Figure 2.3 summarizes these main steps.

It is worth mentioning that process planning is typically an iterative procedure, and it might happen that the design specifications are not compatible with the fabrication technology (see Section 4.1): in this case the control must go back to the design phase for the necessary updates. Fortunately, Additive Manufacturing technologies pose much fewer constraints on the design geometry (wrt to, e.g., CNC milling with limited degrees of freedom), but still there are cases where the aforementioned re-design cannot be avoided.

2.2.2   DESIGN FRAMEWORKS

The papers considered in this book, roughly speaking, all follow a similar design workflow. They begin with the set of capabilities of the manufacturing process, including working volume, material, resolution, and need for support. Together with common and basic rules for additive manufacturing such as connectedness and lack of inner voids, these conceptually define a range of realizable shapes, which is often restricted by individual projects to allow for faster optimization or simpler user interaction.

The next stage of the design process is to formulate a constrained optimization problem, with the constraints ranging from structural soundness to manufacturability, while the objectives encode any of a huge variety of goals. For example, in an architectural context (Section 3.4.2), one goal might be to minimize the number of different types of basic elements or bricks. This reduces the quantity of molds that must be produced, which in turn reduces overall manufacturing costs. For planar-structures design (Section 3.4.1), the important aspect of constructibility is manifested through the ability to assemble the interlocking planes in a collision-free manner. The objectives and constraints are usually provided by or controlled by user interaction, as is some initial guess. The optimization often involves structural, optical, or acoustic simulation, taking the shape as input in either its "primary" representation or some more convenient alternative.

When a final shape emerges that satisfies the constraints and cannot be improved in terms of the objectives, it is passed to a separate piece of code that determines exactly how the fabrication device should be driven (e.g., slices, tool paths, and support), in order to produce a high-quality object with minimum manufacturing time and cost.

Of course, many variants on the above workflow are possible. For example, the details of the manufacturing process could be optimized together with the overall design. The design space could be restricted to a greater or lesser degree, leaving the designer with a lesser or greater degree of control over the ultimate shape. The optimization and simulation could use lower-fidelity versions of the shape, leading to faster computation at the expense of lower accuracy. Initial guesses for the shape, candidate perturbations, or local configurations of materials could be taken from pre-computed or pre-designed libraries. So, while our main focus is on analyzing papers along the two axes of "goals" and "representations," an additional undercurrent throughout the remainder of this book is the variety of ways in which the computational fabrication pipeline can be structured.

The following sections explore the large variety of goals, both general and domain-specific, that have been considered by systems for fabrication-aware design. Before describing those, however, we consider an emerging set of *general* design and fabrication systems with modular or interchangeable parts. These aim at facilitating complex creations for novice, high-level, and professional design, and often allow for the combination of multiple constraints and goals.

One example of a system designed for general-purpose computational fabrication is *openFab*, proposed by Vidimče et al. [2013]. This work defines a fabrication language and programmable pipeline for multi-material additive manufacturing. Inspired by the programmable shaders that form a major part of existing rendering systems, they provide a programmable pipeline to specify volumetric material distribution and thus synthesize 3D geometry for fabrication. To support complex geometries, the pipeline operates in a streaming manner, with only a bounded amount of geometry present in memory at all times. The overall geometry is defined at a coarser scale, and is tessellated to "micro-polygons" on demand. Fine-scale geometric details (e.g., multi-material color dithering, micro-lenses, surface finishing, etc.) are defined procedurally, in a shader-like manner. The framework supports the needs of additive manufacturing, such as: support structure construction (through a 2D grid on the printing plane) and density estimation per printing voxel. Efficient nearest neighbor queries are facilitated through a spatial acceleration structure (BVH).

Instead of manipulating material distribution, the *Spec2Fab* system of Chen et al. [2013a] proposes to describe and design a fabricable object solely by its properties and desired functionality. They provide a framework and API to convert high-level specifications to practical fabrication instructions. The core concept is the combination of a *reducer tree*, in which inner nodes partition the space and leaf nodes assign materials, and a *tuner network*: these are algorithms that are attached to a reducer sub-tree and optimize their parameters. The tuners are a network since they may be interconnected to share information. Examples are combinations of texturing, appearance control (goal-based caustics, shadows, etc.), and deformation behavior. Combined examples produced by the system can be seen in Figure 2.2.

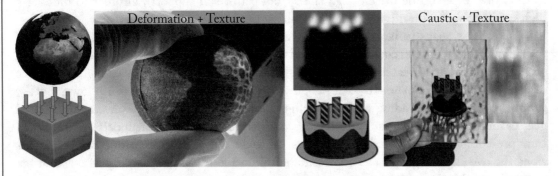

Figure 2.2: 3D printed objects with combined specifications, designed using the system proposed by Chen et al. [2013a]. Left: a miniature Earth model with a prescribed deformation behavior. Right: an optimized surface producing a caustic image as well as casting a prescribed shadow. Images courtesy of Desai Chen.

Clearly, these design frameworks are still preliminary. As this field matures and many different applications and technologies continue to develop, a standardization of the process will most probably be beneficial. Much work and thought is still required, but this approach could facilitate system development and inter-communication between manufacturers, as has been demonstrated many times in the past in other fields (e.g., the previously mentioned rendering, which has seen the standardization of shader stages and languages).

2.2.3   CURRENT PRACTICE FOR PROCESS PLANNING

Modern 3D printing companies and services accept designs in different representations, each leading to a different complexity for the conversion to an effective CAM model. In mechanical engineering, the vast majority of design models comes as a collection of Non-Uniform Rational B-Splines (NURBS) patches with possible trimming curves: such a *nominal geometry* is normally tessellated to form a CAM model and start the process planning phase. When the shape is simple enough, constructive solid geometry (CSG) tools are normally preferred because they guarantee that the resulting model is solid. In this latter case, most CAD softwares provide a tessellation module, although a new trend of tools is emerging to completely avoid the tessellation and perform the whole process planning on the native representation, including the slicing phase. In other areas (e.g., cultural heritage), the input model is usually produced by a 3D digitization campaign which often leads to a triangular mesh: such a representation is already tessellated, although many operations might be necessary to make it actually enclose a printable solid (see Section 4.1).

Whatever the format of the design geometry, the dominating format for the tessellated models used in CAM is the Standard Tessellation Language (STL). STL represents an unstructured collection of triangles but is mostly used to represent structured meshes: thus, the coordinates of any vertex are encoded once for each of the triangles incident at that vertex, which leads to a highly redundant representation. Furthermore, STL files describe only the surface geometry without any representation of color, texture, or other common CAD model attributes. For these reasons, modern standardization efforts (e.g., the AMF format within ISO/ASTM 52915:2013) use indexed representations to avoid redundancy and allow encoding many useful attribute information such as colors, materials, and textures.

Currently, however, typical AM process planning pipelines comprise the following steps.

- *Check and possible adaptation of the input geometry to fabrication requirements.* The (possibly tessellated) geometry must enclose a solid that the target printing technology can actually fabricate (Section 4.1).

- *Building direction.* The model must be correctly oriented to fit the printing chamber, minimize surface roughness and printing time, reduce the need for support structures, etc., (Section 4.3).

- *Creation of support structures.* Depending on both the fabrication tool and the shape, additional geometry may be necessary to support overhanging parts and to keep the part from moving during printing (Section 4.4).

- *Slicing.* The model must be converted to a set of planar slices whose distance might be either constant or adaptive (Section 4.5).

- *Machine instructions.* Each slice must be converted to either a sequence of movements of the fabrication tool (vector-based) or a grid of pixels that define the solid part of the slice (raster-based) (Section 4.6).

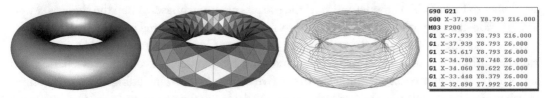

Figure 2.3: Typical process planning phases in additive manufacturing: a design model (left) is tessellated to enter the process planning phase (center-left). Such a tessellation is sliced (center-right), and each slice is converted to a sequence of machine instructions (right). Other operations are typically required depending on the specific technology at hand [Livesu et al., 2017]. Used with permission.

2.3   METRICS/DESIDERATA

Setting up the process planning to fabricate an object is a matter of finding a good tradeoff between different objectives, often depending on the applicative scenario. The PP pipeline can be tuned to strive to optimize for one, or a combination of them. Different criteria have been proposed in the literature. We recap here the most relevant and widely used.

2.3.1   COST

In industrial environments it is quite important to keep the production cost as low as possible. The Generic Cost Model [Alexander et al., 1998] puts together all the variables that control the production cost for a single object, and is designed to be general enough to embrace any layered manufacturing process. Similar cost models have been proposed in [Byun and Lee, 2006a,b, Pham et al., 1999, Thrimurthulu et al., 2004, Xu et al., 1999]. The typical cost model is defined as the sum of three major components: pre-build, build, and post-processing.

Pre-Build Cost
This component measures the cost necessary to turn a design into a set of machine instructions to send to the printer. It also accounts for the low cost (e.g., load the powder into the machine, control or supervise the process planning software) and the time necessary to setup the printer (e.g., cleaning, testing, warming). Methods that aim to minimize the pre-build cost strive for the efficiency of the process, which can be achieved either by using computationally more efficient algorithms or by reducing the user interaction, favoring automatic methods.

Build Cost
Build cost comprises the material cost (for both the desired part and the necessary support structures) and the cost of using the machine for the time necessary to complete the job. The build cost can be reduced in two ways: acting on the printing time or acting on the material waste.

Printing time can be reduced orienting the shape so as to minimize its height (Section 4.3), reducing the number of slices (Section 4.5), or using efficient machine toolpaths (Section 4.6). Material can be reduced by minimizing the support structures' volume, either with a proper choice of the build direction (Section 4.3) or by inserting cavities in the interior of the shape [Lu et al., 2014, Song et al., 2016, Wang et al., 2013]. Recent works aim to reduce material waste and achieve a better surface finish by splitting the part into components that can be printed without support structures (Section 4.2.2). Notice that these methods often trade minimal build cost for the structural strength of the part and, therefore, are not always suitable for industrial production processes.

Post-Processing Cost

The post-processing component measures the labor, material, and time cost necessary to polish the part. This includes: detaching the support structures from the object at the end of the print (Section 4.4.1) and applying some surface finish technique, either manually or through chemical and machine driven processes. These components heavily depend on the process plan. Particularly relevant is the choice of the build direction (Section 4.3) which, in turn, determines the amount and positioning of support structures (Section 4.4.1), and the extent to which the staircase effect introduced by the layered manufacturing process will affect the part quality.

2.3.2 FIDELITY

The fidelity is the degree of exactness with which the part has been reproduced starting from its design. Indeed, layered manufacturing is hardly capable of producing a perfect replica of a given design. Parts of the shape that do not align with the building direction expose a typical *staircase* effect (Figure 2.4a). We distinguish between *form* and *surface roughness*, where the former refers to the overall shape of the prototype, which, to a certain extent, can be quantitatively estimated before printing, and the latter to more local variations of the surface (or high frequencies), which can be approximately estimated only on the printed object since they depend on factors like the printer resolution and the material used. In the following we discuss the most widely used metrics to evaluate the form approximation error introduced by the staircase effect. Being dependent only on process planning parameters, such as the layer thickness and the shape orientation, these metrics are at the core of the algorithms that strive to optimize the PP pipeline (see, e.g., Section 4.3). Afterward, we introduce metrics to evaluate surface roughness, and also discuss some of the practices used in literature to alleviate surface artifacts. Finally, we briefly discuss design compliancy, a criterion that is of fundamental importance in industrial applications, where the printed object is asked to stay within the tolerances set by the designer.

Form

In additive manufacturing objects are created by approximating a given design with a set of layers stacked one on top of the other along the build direction. The goal of form fidelity is to detect

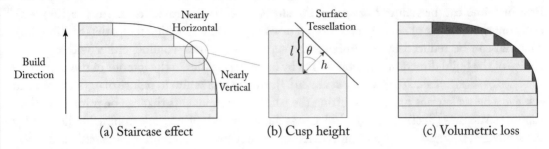

(a) Staircase effect   (b) Cusp height   (c) Volumetric loss

Figure 2.4: Approximating a curved surface with a stack of layers piled along the building direction introduces the typical staircase effect and reduces fidelity (a). As can be noticed, nearly horizontal surfaces introduce more error than nearly vertical ones. Cusp height (b) and volumetric difference (c) are among the most widely used proxies to estimate fidelity [Livesu et al., 2017]. Used with permission.

the difference in shape between these two entities. We present here the two most widely adopted metrics in literature to evaluate the error introduced by layered manufacturing: the cusp height error and the volumetric difference.

- The *cusp-height error* concept was introduced in the context of adaptive slicing for layered manufacturing [Dolenc and Mäkelä, 1994]. It is defined as the maximum distance between the manufactured part's surface and the design surface [Alexander et al., 1998]. It depends on the layer thickness l and the angle θ between the local surface orientation and the build direction (Figure 2.4b), namely

$$
h = \begin{cases} l|\cos\theta| & \textbf{for} \quad |\cos\theta| \neq 1 \\ 0 & \textbf{for} \quad |\cos\theta| = 1. \end{cases}
\tag{2.1}
$$

Note that $|\cos\theta|$ is very small for nearly orthogonal angles and grows up to 1 when θ is close to 0 degrees. This well encodes the big approximation error difference between nearly vertical and nearly horizontal surfaces, as shown in Figure 2.4a. The integral of the cusp height on the whole surface is a good estimator of the fidelity of the printed model. To this end, for triangulated surfaces the cusp height is computed separately on each facet as the dot product between the build direction $\mathbf{b}$ and the triangle normal $\mathbf{n}$, $|\cos\theta| = |\mathbf{b} \cdot \mathbf{n}|$. It is then scaled by the triangle area and normalized by the total mesh area so as to accommodate uneven tessellations and make it scale independent [Wang et al., 2016].

- The *volumetric difference* is the difference between the volume enclosed by the design and the volume of the printed object (red area in Figure 2.4c). Masood et al. [2000] explain how to evaluate the volumetric error for simple geometries such as cylinders, cubes, pyramids and spheres. Such work was then extended to complex geometries in Masood et al. [2003].

Notice that, although they are both used as proxies to minimize the staircase effect, volumetric difference and cusp height are not equivalent. Taufik and Jain [2014] express a preference for volumetric error. They observe that, if the shape to be printed contains steep slopes, to a little variation in cusp height may correspond a large variation in volumetric difference, thus making the latter more accurate than the former.

Surface Roughness

Also known as *surface texture*, or *surface finish*, roughness aims at measuring tiny local variations on the surface, which determine visual (e.g., the way the object reflects the light), haptic (e.g., the porosity of the surface), and mechanical (e.g., friction) properties of the shape. Unlike form, surface roughness cannot be easily estimated prior printing, as it mostly depends on parameters like the printer resolution and the printing material. Usually, it is measured directly on the printed object, by using sampling techniques [Townsend et al., 2016].

In industrial environments the design specifies both the desired surface roughness and the metric that should be used to estimate it. The arithmetic mean surface roughness (R_a) is by far the most widely adopted metric, as stated in Townsend et al. [2016]. Delfs et al. [2016] observe that the surface roughness depth (R_z) is a better proxy to measure surface finish, as it is well representative of how a human eye assesses surface quality. Both R_a and R_z are roughness metrics defined in the ISO 4287 standard [ISO, 1997]. They are computed on profile curves obtained by cutting the surface with an orthogonal plane. We point the reader to Mitutoyo [2009] for a nice explanation of how these metrics are defined and can be estimated in practice; much other information can be found online.

Different strategies have been proposed in the literature to achieve the best surface quality possible. Reeves and Cobb [1997] evaluate the *meniscus smoothing* to alleviate the impact of the staircase effect in stereolithography and produce smoother surfaces. It consists of an edited build cycle in which each layer, after solidification, is lift above the upper surface of the resin tank to stretch a meniscus of liquid between each polymerized layer. The resin meniscus is then solidified by using scan data from the previous layer, producing a smoother transition between adjacent layers. In the case of plastic deposition technologies (e.g., FDM) acetone vapor baths are commonly used to obtain a smoothing effect on the printed object [Lalehpour et al., 2017]. Other authors have recently observed that additional artifacts that affect surface finish may be introduced while detaching support structures from the object. In fact, tiny features may be too weak and break during this process, leaving residual support material attached to the surface. Zhang et al. [2015b] propose a perceptual method that optimizes for the location of the touching points between surface and supports. Their system tries to *hide* support removal artifacts by placing them at the least salient parts of the shape, as far as possible from its perceptually relevant features. In general, the extent to which support removal may affect surface quality depends on the printer and the material used. In metal printing supports removal is extremely challenging due to the properties of the material involved. Some recent methods to alleviate this problem are discussed in Section 4.4.1.

Design Compliancy

In industrial AM environments object compliance with the original design is requested, meaning that both the form and the surface roughness must stay within precise error bounds set by the designer. To this end, a number of form and orientation tolerances are often used in industrial design. Typical form requirements regard the straightness, planarity, circularity, or cylindricity of the components. Regarding orientation, typical requirements are parallelism, orthogonality, or angularity. Notice that, similarly to roughness, these tolerances can only be estimated after printing. Metrology is a vast field, and we do not discuss details regarding how these quantities can be estimated. We point the reader to Huang et al. [2014b, 2015], Jin et al. [2015], Townsend et al. [2016] for further details on this topic.

2.3.3   FUNCTIONALITY

When the object to be printed is the result of a shape optimization process, it is typically asked to meet some prescribed functionality requirements. We consider three broad categories: (1) requirements on the robustness of the shape, such as resiliency with respect to previously known or unknown external forces; (2) requirements on the mass distribution, for example to achieve static or dynamic equilibrium; and (3) requirements on thermal and mechanical properties, for example regarding the heat dissipation or the stiffness of industrial components.

Structural Soundness/Robustness

With 3D printing consumers can directly produce their own objects, but not always the digital shapes they started with were meant to be fabricated, and can therefore reveal to be excessively fragile and easily break under cleaning, transportation or handling. Zhou et al. [2013] introduced the worst-case structural analysis to detect the most fragile parts of an object (see Figure 2.5). In Langlois et al. [2016], a stochastic finite element method to compute failure probabilities is presented. A number of methods [Li et al., 2015a, Stava et al., 2012, Xu et al., 2016] combine a lightweight structural analysis with automatic systems that enforce weak features through operations such as hollowing, thickening, strut insertion, and inner structures.

Mass Distribution

The distribution of material and cavities inside a 3D printed model has been the subject of recent research. In Prévost et al. [2013], a method to optimize the balance of a shape to make it stand in a given pose is proposed. The work was further extended in Bächer et al. [2014], where a novel optimization of the mass distribution to make an object spinnable around a given axis was proposed. The optimization of the buoyant equilibrium is the subject of Wang and Whiting

Figure 2.5: Structural analysis to detect weak object parts from Zhou et al. [2013]. Used with permission.

[2016], where a method to create floating objects in a prescribed pose is presented. In Prévost et al. [2016], the authors consider the standing, suspension, and immersion balancing problems for 3D printed objects containing embedded movable masses. In Musialski et al. [2015], Wu et al. [2016a], unified frameworks for the optimal interior design and mass distribution of 3D printed objects that enable the optimization of static, rotational, and buoyant equilibrium are presented. Recent research has pushed this concept even further. In Musialski et al. [2016b], the shape design is posed as non-linear problem that aims to optimize the natural frequencies of a shape, for example to make it sound in a controlled manner, as a musical instrument.

Thermal/Mechanical Properties

Additive manufacturing enables the fabrication of shapes that would be impossible to produce with classical subtractive techniques. Shapes that in the past were interesting only from a pure theoretical standpoint can now be printed and their functionality exploited. To this end, additive manufacturing has fostered a lot of research in fields like topology optimization [Brackett et al., 2011, Dede et al., 2015], where the goal is to generate shapes which optimize performances in terms of some physical requirement (e.g., weight, heat dissipation, stiffness). This is important in fields like aerospace industry, where the size and weight of components is a crucial factor, or for the generation of injection moulds, where optimized cooling systems may increase both the productivity and the overall quality of industrial products. Alongside, researchers have been investigating the correlation between the structure of a shape and its physical properties. Recent works aim to optimize the local structure of a printed object to produce controlled deformations or to meet precise rigidity requirements. We review this body of literature in Section 4.4.2.

CHAPTER 3

Design for Additive Manufacturing

3.1 APPEARANCE

We now turn to describing fabrication-aware design systems that optimize for specific goals, beginning with appearance. For many applications, such as furniture or toy design, an object's appearance is as important as its physical or functional properties. Of course, the field of computer graphics has extensively studied the way objects look, and hence has naturally addressed these needs also in the fabrication context. In this section, we will describe fabrication-aware design aimed at shape appearance, both in the optical sense (Section 3.1.1) and geometric sense (Section 3.1.2).

3.1.1 LIGHT INTERACTION

Studying the behavior of light has been at the heart of computer graphics ever since the field was created. In the past decade, the techniques developed in this field have been adapted to tackle the problem of controlling the interaction of physical objects with light, through computational fabrication. Note that we mention here only some of the most recent work, concentrating on additive manufacturing applications. A lot of work has been done to control shape appearance through optical phenomena such as shadows [Baran et al., 2012, Bermano et al., 2012, Zhao et al., 2016b] or detailed Lambertian texturing [Panozzo et al., 2015, Schüller et al., 2016, Zhang et al., 2015d]. We refer the reader to focused surveys for more details on other appearance fabrication applications (such as more elaborate reflection distribution function, or BRDF, control) [Hullin et al., 2013].

An immediate need that arose from multi-material printing technologies is approximating desired appearance through spatial mixing of different colored materials. Hergel and Lefebvre [2014] proposed a method to improve the quality of such multi-color prints in the case of multi-filament extrusion, by careful planning of the printing paths. First, they optimize the azimuth (angle around the z axis) of the object to reduce the chance of one extruder smearing while the other is active. The intersection volume induced by the paths of the two extruders is approximated based on Boolean mesh operations, and the best angle is determined through an exhaustive search. Second, they build a disposable wall (rampart) in proximity to the object, in the form of an offset surface, and thus creating a cleaning station which does not require long travel times. Finally, they alter the path planning to avoid smear and oozing when extruders are idle or traveling over the object. 2D slices with contour traces are considered for path planning. Different parts of the slice are printed according to ambient occlusion computations (on a sparse 3D grid), to hide path interfaces in less visible regions.

A different approach for better color control is via halftoning. Reiner et al. [2014] propose to slightly modulate the printed shape in a different manner on every slice. In low-end devices, this will expose or hide material of certain colors and thus minimize effects that are due to low accuracy. They propose a sine function modulation of the object's surface to offer smooth gray-level control for black and white material prints. For translucent materials, on the other hand, another method was introduced. This method determines color values for each voxel close enough to the surface, taking into account the printer's tonal gamut and subsurface scattering [Brunton et al., 2015]. The method diffuses the error locally following a novel iso-distance surface traversal scheme, and produces the results on fixed sized batches of slices (implying bounded memory and processing time). The desired appearance is converted to CMYK tonal values, and the error is diffused using a function similar to traditional 2D halftoning.

Of course, light interacts with manufactured surfaces in the entire volume, and not just near the surface. Papas et al. [2013] proposed a method to replicate the appearance of translucent materials using pigmented silicone. A set of silicon pigments is measured using a custom-designed spectrometer, and their translucency profile is estimated for five representative spectra. The same process is done for a target material, and a recipe for pigment mixing reproducing the appearance of the given material is sought. The optimization considers both color and reflection profile, making the fabricated result resemble the target under different lighting conditions and geometric shapes. Later, a conceptually similar analysis was performed for 3D printing resins, enabling more accurate reproduction of high-frequency color details during 3D printing, enabled through proposing a novel optimization operating over a Monte Carlo simulation [Elek et al., 2017].

Another approach to controlling the volumetric propagation of light is that of Pereira et al. [2014], who studied 3D printing with multiple transparent materials in a way that yielded collections of optical fibers embedded in an object. Given a shape and its parameterization, a routing optimization computes fiber paths within the shape that conform to fabrication constraints and target light routing from a given plane to its corresponding points on the mesh's parameterized surface. Fiber fabrication is done by printing a clear material, coated by a low index-of-refraction material (support material in this case). The fibers within the shape are optimized to be as smooth and as far apart as possible, to maximize internal reflections. The solution is formulated using implicit functions, such that each fiber is traced out by level sets. Later, the same group has also designed a method to control anisotropic specularity effects (or, more broadly, BRDFs), by embedding magnetic flakes during the printing process [Pereira et al., 2017]. The printing is performed with a curable resin, mixed with metallic flakes, in a controlled magnetic field. The field aligns the embedded flakes in desired directions, thus controlling their reflectance direction. After a layer has cured, the flakes are not free to move anymore, allowing the magnetic field to be changed according to the desired effect of the next layer.

Because of the great interest in light interaction within graphics, it is natural that a large number of fabrication-related papers in the community have addressed this. The underlying representations have involved various subsets of the plenoptic and scattering functions (see Hullin et al. [2013] for details), often with novel constraints imposed on the optimization to reduce computational cost. In all, we are remarkably close to being able to reproduce many components of objects' appearance in a way that is indistinguishable from the originals: much as graphics has slowly approached photorealism in rendering, photorealism in 3D fabrication now appears to be within reach.

3.1.2   GEOMETRIC PATTERNS

Enriching the visual aspect of designs is not restricted to appearance alterations alone, but can also be done by adding intricate structural details. In the case of fabrication-aware design, an interesting challenge is to design objects consisting *only* of these patterns, instead of having a base object that is merely augmented by them. This is challenging due to the need to satisfy manufacturability constraints, while keeping the shape as similar as possible to the target.

To address this, a method for topology-constrained pattern synthesis was proposed by Zhou et al. [2014a]. Given a 2D exemplar, the method strives to tile it along a curve in a fabricable way. The exemplar is divided into several vertical pieces, which are allowed to be connected in any order, as long as they have the same quantity of "portals" (non-void regions) along their interface. A dynamic program is employed to find the best combination of the sub-pieces, consisting of parts as whole as possible while maintaining connectedness. Other parameters are optional, such as the number of holes. Since drifts are introduced when connecting out-of-order sub-parts, a smooth deformation aligns the result back to the given target curve in a final step.

Later, Martínez et al. [2015a] extended this approach to combine texture synthesis with topology optimization, i.e., to optimize for compliance under given loads and for similarity to a given example (see Figure 3.1). This method is designed for the 2D case, but it should be extendible to 3D with some effort. This method divides the plane into grid elements, which

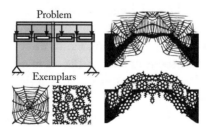

Figure 3.1: Example-based topology optimization [Martínez et al., 2015a].

are assigned void or full labels by the end of the process. Compliance is formulated by having an objective that combines constraints computed by the Finite Elements Method (FEM), and local shape similarity with respect to the exemplar (as is popular in texture synthesis techniques). The highly nonlinear problem is solved through the Globally Convergent Method of Moving Asymptotes (GCMMA).

In 3D, the approach of choice for this problem is covering the given surface with the exemplars. Dumas et al. [2015] propose a method adapted from traditional by-example texture synthesis. Given a mesh and a pattern image, the method carves the mesh out. The carving seeks to resemble the input patterns, while respecting fabricability, in the form of connectedness, as well as resistance to given external and internal forces. The shell is hierarchically voxelized (resembling an octree), and each voxel is projected onto a plane. The planes are copies of the input image, distributed on the unit sphere. Image directions and positions on the sphere are determined by optimizing smoothness of image directions and colors. The carving is determined according to the color of the projected positions. After carving, a simplified version of the mesh and its new connectivity is constructed, from which an approximative shell is created for FEM simulation. When weak spots are detected, neighboring pieces are connected through "bridges" to reinforce the mesh, and the process is iterated to incorporate these additions in a pleasing manner.

Chen et al. [2016] propose another solution for the same problem. The input basic exemplar elements are represented by their medial axes, and are distributed on the surface using a blue noise method. These are then translated and deformed to improve visual quality, aiming at tangential connections or partial overlaps between elements. This step employs the Hausdorff distance for shape matching and a connectivity graph to ensure rigidity. After the optimization, the medial axis is thickened to become a volumetric mesh for FEM analysis. If weak spots are found, new elements are added and the previous steps are repeated until convergence. The resulting patterns, and their potential applications are depicted in Figure 3.2. These elements could also be printed on a flat plane in patches of connected components, and be curved up and assembled together manually afterwards, as proposed in a later work [Chen et al., 2017b]. This process requires a manual step, but saves printing time, support material, and increases the maximum allowed volume of the object.

Zehnder et al. [2016] introduced an interactive way to cover a surface with user defined curves. The curves are drawn in 2D, and are represented as b-splines. Then, they are lifted to 3D and are embedded on the surface with minimal distortion. The shape is represented as a Loop subdivision surface, and elastic rules are applied to the curves in order to maintain their shape (geodesic curvature) and fit them to that of the surface (normal curvature). The curves can be manually placed, or randomly distributed and grown (in a user-controlled fashion) until reaching each other. Structural analysis is then run for shape stability and avoidance of large local deformations. Modal analysis (see Section 3.2.1) is used to find load vectors, and virtual

Figure 3.2: Fascinating and aesthetic appearance can be created by modulating non-detailed objects (right), with combinations of pattern exemplars (left), as presented by Chen et al. [2016]. Used with permission.

edges are added between the curves to measure the amount of deformation. Large deformations of virtual curves are pointed out to the user for correction (by adding elements).

Repetitive patterning has been used to decorate both functional and ornamental objects for centuries. As seen in this section, additive manufacturing has enabled the design of shapes consisting of the patterns alone, without a supporting base surface. In addition, the methods here enable this type of design for novice users, which until now was only in the realm of expert jewelers and woodworkers. The analogy to texturing and texture synthesis is straightforward, and hence parameterization is naturally employed by these methods. In the future, it would be interesting to combine such patterning techniques with applications directed more at structural integrity, such as decorative yet functional furniture.

3.2   STRENGTH AND MOTION

In addition to appearance, many of the fields studied by the computer graphics community are applicable to computational fabrication. Subjects such as shape analysis, deformation, animation, high-level goal-based optimization, and physically based simulation are some of these areas of expertise, and all have been applied to exciting new manufacturing applications. As we will see in this section, adapting traditional computer graphics knowledge to fabrication-related applications enables addressing new problems that were not considered in the past, especially relating to deformation and motion. Efficient physically based modeling for complex geometries has been researched in computer graphics for many years, yielding different models and representations [Nealen et al., 2006]. It turns out that these techniques are mature enough to approximate real-life physical objects, enabling applications such as meta-material and moving character design, when combined with example-based modeling or animation (Sections 3.2.2, 3.2.3, and 3.2.4); however, we start with a classic aspect of any physical design—structural analysis (Section 3.2.1).

3.2.1   INTEGRITY

The most prominent aspect of fabrication design is structural integrity analysis, i.e., making sure a design is durable under both its own weight (dead load) and other prescribed forces (live load). As we will see in later sections, structural analysis is an integral building block for most fabrication-aware tasks. An object's strength is classically analyzed by simulating the amount of deformation (or *strain*) the volumetric body undergoes when external and internal forces are applied to it. Typically, the user specifies the amount and location of said forces, and an FEM-based simulation is performed. Areas that result in excessive deformation are considered as weak points, which must be eliminated with revisions to the design. While this method yields accurate results in the vast majority of cases, it is sometimes impractical because of several reasons. Running a simulation correctly requires expertise that are not common to novice users and artistic designers. Specifying external stimuli for everyday usage of various objects can be an ill-posed or unfeasible problem. Simulation runtimes can also be tedious, especially when many iterations and experiments are needed. These challenges and others motivate this highly active field of research.

A possible solution for some of the mentioned challenges is to employ modal analysis [Zhou et al., 2013]. Modal analysis is traditionally used for vibration validation, but by testing several modes (which correlate to several vibration frequencies) one obtains a good guess for consistent weak spots. Such an analysis can locate structural weakness without any specified forces, and hence is a worst-case structural analysis, which nevertheless can easily be performed by novice users. In this method, a given volume (represented by tetrahedra or a triangular mesh that is tetrahedralized) undergoes inspection of the spectra and eigenfunctions (or modals) of its Laplacian operator. More recent work [Langlois et al., 2016] has employed stochastic Finite Element analysis (or *SFEM*) for the same problem, analyzing the probability of an object to fail based on reasonable usage, rather then worst-case one. The method asks the user for a typical use case (e.g., the object is dropped on the floor), and a stochastic process estimates the loads that the object undergoes. To improve performance, an SVD-based dimension reduction on the force distribution is proposed. In addition, the effectiveness of this robustness analysis is demonstrated by using it to guide a topology optimization process. Another solution, proposed by Umetani and Schmidt [2013] to significantly improve the performance of integrity analysis, is to consider only cross-sections of the volume. The method is able to analyze the structural integrity at interactive rates due to the extension of the Euler-Bernoulli assumption to 3D meshes.

The latter assumption is common practice in beam engineering, where the moment arm is much longer than the cross-sectional dimensions. The method identifies, clusters, and analyzes such cross-sections in the object, and applies only to thin elongated parts.

Of course, once analysis has identified weaknesses, the structural integrity of the designed object can also be automatically improved. Stava et al. [2012] proposed a system for analysis and correction of structural integrity for general 3D objects. The analysis is done by traditional FEM simulation on the tetrahedralized interior of the shape. More unusually, the external stimuli are computed automatically by finding pinching pairs positioned where the objects are most likely to be grasped. The correction step considers the medial axis of the shape for improved performance and simplicity. Three types of corrections are proposed: adding struts, hollowing the interior, and thickening. The different options are prioritized according to their benefit and estimated aesthetic effect, and are iteratively proposed to the user.

Hollowing the volume is an effective way to reduce material and weight while maintaining the same outside shape. Therefore, it has been shown that hollowing in a honeycomb-like manner is very effective in terms of material-to-strength ratio [Lu et al., 2014]. In this method, the internal volume of a given mesh is tessellated and the stress is analyzed under given external loads. Then, sites are distributed inside the volume according to the computed stress map (in a halftoning-like manner), and Voronoi cells are created. Each cell is voxelized and a harmonic distance function is constructed in it. The cell is then hollowed according to the iso-surfaces induced by the distance function at a user-given value. This process, also known as porous extraction, ensures a smooth structure with controllable thickness.

Instead of hollowing, topology optimization has also been employed to improve the material-to-strength ratio, while remaining as close as possible to a prescribed shape. Christiansen et al. [2015] propose a combined shape and topology optimization approach, where an initial shape is optimized to have a prescribed volume and withstand boundary conditions (loads). This work develops a topology optimization process that tries to remain as close as possible to the initial state. The initial meshed object, along with a box around it, are tetrahedralized, and an optimization is run to minimize the combined shape and FEM-based stress objective. The optimization is run on vertex positions of the tetrahedra and their labels (which can be solid or void). For numerical stability, a good tessellation of the volume is preserved throughout the process by remeshing operations performed after deformation iterations.

Wu et al. [2015] maximize strength at a prescribed material consumption ratio using topology optimization. Given a volume, it is first decomposed into a regular hexahedral grid. The stress distribution is analyzed, and the topology, only inside the volume, is evolved, in alternating steps. Through an efficient GPU implementation, this method is designed to accommodate high-resolution shapes with millions of elements. The topology optimization process is also altered to incorporate fabrication-related constraints such as minimum thickness, and design considerations such as symmetries and pattern repetition. Figure 3.3 depicts two fabricated results of the system for specified loads.

Figure 3.3: Example result from the method introduced by Wu et al. [2015]. Both models were discretized with several million elements, and optimized in a few minutes. Used with permission.

Efficient integrity analysis and optimization still encompass many open research questions. Accurate analysis for complex shapes at interactive rates will reduce design time significantly and potentially allow a process free of integrity concerns for the designer. To achieve this, correct approximations and reductions most probably need to be made, similar to the previously described use of modal analysis and medial axes.

3.2.2   COMPLIANCE

Elastic deformations are a fundamental part of the physical behavior of objects in our everyday life. The deformation of materials according to applied forces depends on many factors, from the molecular level to global object geometry. However, it is well known that the behavior of materials with micro-scale inhomogeneities can be uniformly approximated at medium scale, through numerical coarsening and *homogenization* [Kharevych et al., 2009]. Controlling the deformation behavior of materials through design of their micro-scale structures (called hierarchical materials, or *meta-materials*) is traditionally in the realm of material sciences [Kalidindi, 2015].

Pioneering work in material design in the field of computer graphics employed example-based modeling, a well established approach in the field [Bickel et al., 2010]. The goal of the project is to 3D print materials with desired deformation behavior. In order to do so, first a set of base materials is collected and the deformation of each is measured using a robotic system. From these observations, via a homogenization algorithm, the nonlinear relationship between stress and strain is deduced to enable FEM simulation. This allows prediction of the material deformation not only for different geometries, but also for different combinations of stacked materials. The approximation is done using radial basis function (RBF) interpolation, and the desired deformation is achieved through layering of the different materials. The target object is divided into voxels, where each can be assigned a single material. The material assignment problem is solved via a branch-and-bound algorithm with clustering. Quasi-static FEM is used to simulate the object and test whether the material assignments fit the desired behavior. The system and its results are illustrated in Figure 3.4. Later, the same approach of example-based modeling was used to design a silicone patch that forms specific wrinkles in the context of facial animatronics [Bickel et al., 2012]. In this case, only one material is used and only its thickness is optimized (due to manufacturing constraints). Similarly, material properties are studied through example-based vision of deformations, but in this context continuum mechanics must be employed due to the large deformations. All of these methods, however, assume limited deformation and are inaccurate during large deformation due to the linearization process involved. Using a higher-order element can help mitigate these inaccuracies, at the cost of computation and difficult parameter tuning. To this end, a method was recently proposed which is able to perform numerical coarsening over a shape and still remains very accurate for highly dynamic objects which present large scale deformation. Coarsening reduces the number of elements required during the simulation. The proposed method is able to do that while achieving more accurate results with a speedup of between one and two orders of magnitude. The method starts with a coarse hexahedral mesh and subdivides it as necessary in order to match example data. In addition, a new method to handle impact-response is also proposed, to complement the accuracy of the material simulation. The accuracy and performance are demonstrated through the design of jumping objects, which achieve the desired jumping height and landing orientation [Chen et al., 2017a].

Xu et al. [2015] presented an interactive material design system for prescribed deformation under given external forces. The material distribution (density) is optimized continuously, and converted to discrete, fabricable material assignments in a later step. To speed up the optimization, the eigenvectors of the mesh Laplacian are computed, and the reduced model is optimized, instead of all tetrahedra directly—this approach is called modal reduction, and is similar to the modal analysis mentioned in Section 3.2.1.

Another approach to material design is through *micro-structures*, i.e., precomputed elements with known deformation behavior that are optimized to be tiled together to reproduce a target property. Panetta et al. [2015] introduce a collection of tilable and fabricable base elements, spanning various combinations of Young's moduli and Poisson's ratios. A constrained exhaustive search over basic elements is performed to map out the different tile families and their properties. The search is constrained to tilable, manufacturable, and isotropic elements, and is performed by changing edge thickness and connectivity over 15 vertices inside one tetrahedron (which is reflected cubically to achieve isotropy). For various examples of given target shapes and mechanical properties, an object is fabricated through lithography, combining the different tiles to control the deformation behavior. In follow-up work, the range of achievable mechanical properties is widened through a novel optimization scheme, and worst-case structural integrity is also considered [Panetta et al., 2017]. Similarly, Schumacher et al. [2015] use spatially

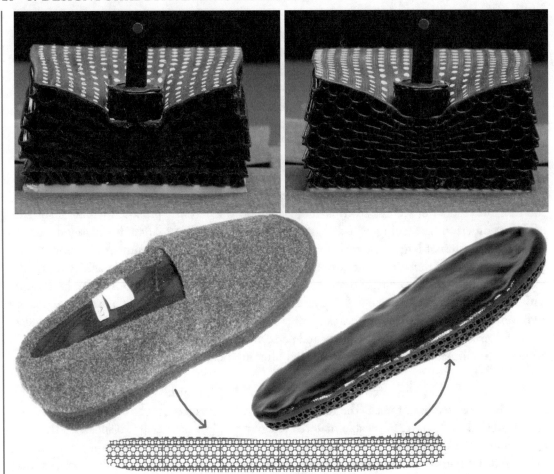

Figure 3.4: Data-driven material modeling, as performed by Bickel et al. [2010]. Top: the deformation behavior of base materials and micro-structures is measured in a robotic arm system. Bottom: these are then stacked to replicate the deformation behavior of other measured objects. Used with permission.

varying micro-structure to control elastic properties. A database of micro-structure families is constructed through topology optimization on small voxelized cells, with mechanical properties computed through numerical coarsening. Then, a distance function is generated to enable interpolation between the precomputed elements during the design step. Unlike traditional topology design approaches, connectivity between cells is considered, and so are multiple candidates (instead of just one) during the optimization. This gives the optimization more flexibility in changing already-computed regions to achieve the target deformation response regardless of target shape, all while maintaining manufacturability. These types of micro-structures can also be used as building blocks to a topology optimization process, allowing it to handle larger volumes than traditional voxel-based methods, as demonstrated by Zhu et al. [2017].

Recently, a stochastic approach to generate elasticity-controlling micro-structures in a streamlined manner was proposed [Martínez et al., 2016]. In this setting, the elasticity (Young's modulus) of printed objects is controlled by varying the density of the micro-structures, following an open foam structure. The volume is sampled according to the desired density (derived from the desired Young's moduli through homogenization), and Voronoi cells are created accordingly. The edges of these cells constitute the open foam, and these are generated only on demand in small chunks, to enable streamlining. This construction ensures printability (no enclosed voids, and one connected component) and allows natural gradations of elasticity without any global optimization, both for the discussed isotropic case, and as later proposed, for the orthotropic one [Martínez et al., 2017].

Deformable material design potentially allows the use of less material for elaborate deformation response, and could be combined with any of the other fields in computational fabrication to improve efficiency and performance. The work described in this part heavily depends on manufacturing technologies, and will most likely continue to evolve as additive manufacturing technologies mature. The proposed solutions focus mainly on explicit representations, enabling FEM simulation. However, the use of other representations, such as modal decomposition and Voronoi diagrams, are intriguing solutions for complexity reduction.

3.2.3 DEFORMATION

The same elasticity properties that are used in the previous section for material design could also be used to accurately predict the shape of a fabricated piece under given forces. This facilitates designing objects that match a desired shape, or several shapes, under different applied loads. In such cases, the inverse problem to the classic forward simulation must be solved, i.e., an initial shape is sought such that it will resemble a target *after* it is deformed by prescribed forces. For example, consider printing a shape that consists of two bulky parts that are weakly connected. Such a shape will probably deform under gravity. However, this can be compensated for, resulting in a fabricated shape with a distorted connection that is correct after the aforementioned deformation. This example can be challenging since classical iterative methods tend to converge very slowly due to the low forces involved. It has been shown that ANM (Asymptotic Numerical Method) can be exploited to drastically improve performance in such cases [Chen et al., 2014]. As another example, Skouras et al. [2012] developed a physical model and an optimization system to compute the needed rest shape of an inflatable balloon such that it will match a target shape once inflated. An interesting usage for such approaches has been proposed in the context of reusable molds [Malomo et al., 2016]. Under this paradigm, a mold, which is traditionally either rigid or usable only once, is created from a flexible material, and should support opening without tearing. Given a shape, its outer shell is trivially computed, which is decomposed using a normal direction clustering algorithm. These clusters could already be used as a mold with many connected components. To reduce the number of components, and the number and length of cuts within a component, the clusters are merged together in a greedy manner. When removing a cut (merging two adjacent clusters), the new mold's feasibility is examined using a simulation with heuristically generated forces, which are trying to detach the mold from the object. The simulation is done using a projective dynamics method, and a thin-shell representation, where the clusters are re-meshed for numerical quality.

The same goal, of matching a shape after deformation, can be applied for several target shapes under different boundary conditions. This will pose similar computational challenges. Pérez et al. [2015] propose using a rod network to approximate the shape and reduce the dimensionality of the problem. A given triangular mesh is converted to hexagons through Voronoi tessellation, and the shape is approximated by fabricating only their edges (which are more compliant to stretch and shear). Using a rod representation (having reduced coordinates of just the central line and angles) with an elliptic cross section, the method optimizes the major and minor radii of the rods to match the desired target shapes. By correctly choosing two orthogonal directions of the ellipse, the optimization is able to control in-plane and out-of-plane deformations independently. In follow-up work, the same mechanism is used to design Kirchoff–Plateau surfaces, i.e., rods are 3D printed on a pre-stretched membrane, which deforms into its steady state upon release. The rods are used to control said steady state. Methods to simulate the steady state of the membrane under the designed rods in real-time, and to alter the rod design in order to achieve desired deformations in the steady state space are proposed, based on the same in-plane and out-of-plane control. A similar idea was concurrently proposed in Guseinov et al. [2017], where shapes are manufactured from two pre-stretched elastic sheets and 3D printed elements which are glued to them. Given a desired shape, it is uniformly triangulated and each triangle is assigned a 3D primitive in the optimization. The optimization dictates the 3D shape and position of these frustum plus attachment pins primitives so that when glued to the pre-stretched sheets they will force a desired steady state upon release.

A more elaborate process can include multi-materials and the optimization of the actuation forces themselves. Such a system was suggested in the context character design [Skouras et al., 2013]. The input is a rest pose and a set of target shapes, and the system determines the volumetric material distribution, along with the actuation system (actuating forces, locations, and quantity). The method first distributes many actuators along the object's boundary, and computes the needed forces for each pose, restricted to physically viable directions, depending on the actuation type (manual posing, pin-based, or string-based—see Figure 3.5, courtesy of Skouras et al. [2013]). In a second step the actuators are unified according to a sparsity parameter, and lastly the material distribution is determined to fine tune the poses once the actuation is set. Another system involving multi-material and the actuation system was proposed in the context of soft pneumatic deformation [Ma et al., 2017]. A complete pipeline to design and fabricate objects which deform with internal air chambers is proposed. Given the target rest pose and deformations, the shape is tetrahedralized and regions with similar deformations are clustered together. Each cluster is then considered as a chamber, and its boundary is quadrangulated. The chambers are skin-frame structures, where the quadmesh is a frame, designed to orient the isotropic pneumatic deformation. The stiffness of different parts of the frame is optimized, using different materials during the print to achieve the target deformation, while the rest of the chamber boundary is printed with a flexible material.

Figure 3.5: An example from Skouras et al. [2013]. Used with permission.

An exotic example of deformable design is that of tensegrities: networks of rigid and elastic elements that are in static equilibrium [Gauge et al., 2014]. Since these states of equilibrium are difficult to achieve, the proposed method requires a predefined library of stable building blocks that may be altered. During the design process the user can connect the basic library elements and move individual vertices, while an optimization process determines the length of required cables to maintain the equilibrium during these changes. Later, Pietroni et al. [2017] devised a more elaborate system able to convert given shapes to tensegrities. The key insight is to use geometric conditions, which are more restrictive (but are shown to be reasonable) than an actual physical simulation, to determine stability of the structure. Given a mesh, a dense graph connects all of its vertices to themselves. Then, an alternating discrete/continuous optimization chooses which edges should be struts and how to move vertex positions in order to get a stable configuration. Last, redundant cables are removed.

In summary, deformable shape design has experienced great progress thanks to additive manufacturing. Processes that involved weeks of trial and error by experienced professionals can now be dramatically sped up through simulation and accurate fabrication. In the future, soft and deformable characters are likely to be dominant in physical character interaction with humans. A purely soft character would pose no risk to a human, even when malfunctioning. In order to realize such characters, however, novel actuation mechanisms would probably need to be conceived.

3.2.4   MOTION

The design of rigid bodies, connected with joints, has received a lot of attention recently. Such objects can be figurines or serve a supporting function when they are posable. More elaborate versions, which include mechanisms such as gear systems and motors, present a wide range of movement and can even be used to build walking robots. Realizing moving digital models physically in an automatic and efficient way is one of the holy grails of computational fabrication. This dream, of having a 3D counterpart to the traditional 2D print button in design tools, both for CAD and animation, still requires much work; however, creating such objects via assemblies, rather than all at once, is still a significant step on this path.

One-piece fabricated models that can stably and independently hold different poses were proposed concurrently by two different publications. Cali et al. [2012] proposed a system to design articulated models. A natural input for such goals are rigs [Sturman, 1998], which define the object's range of motion and thus already include locations of possible joints. The proposed method converts each such position in the rig to a mechanical printable joint from a predefined library. The user can specify angle constraints on the joints, and the method automatically adjusts the results to be fabricable, collision-free, and withstand gravity. In concurrent work, articulated characters were fabricated from skinned meshes [Bächer et al., 2012]. Skinning is the process of controlling deformations of a given object using, typically, a skeleton. The surface (or *skin*) deforms along with the skeletal bones, according to the assigned weights [Jacobson et al., 2014]. In this case, the skinning weights are used to compute candidate joints (from a library), and are placed on the approximated medial axis. Structural integrity is approximated by considering cross-sectional width at joint locations. Recently, a method to convert rigidly articulated joints in assemblies to ones that are based on compliant mechanisms has been proposed [Megaro et al., 2017]. Compliant mechanisms are gaining popularity in engineering fields since they are more elegant, robust, and efficient, since they can be manufactured as one piece. Choosing from a catalog of predefined parametric compliant mechanisms, the proposed method automatically replaces rigid joints wherever possible while validating lateral stability, actuation range, avoiding collisions, and, perhaps most importantly, preventing material failure. Another interesting articulation mechanism that was recently explored is based on telescoping structures [Yu et al., 2017]. This novel mechanism allows for organic movement and compact storage. The proposed method develops a new parametrization to support the feasibility of such structures, and shows how one can generate them along a G1 helical curve. From this insight, a dictionary of possible basic elements is developed. Then, an interactive design tool for such structures is presented. Using shape skeletonization, the framework is able to combine several telescoping curves in a tree-like manner, significantly enlarging the space of designable shapes.

In addition, more elaborate rigid motion can also be carried out through mechanical elements, such as pulleys, gears or sliders. Such *automata* can be driven by a motor or by hand, and their design traditionally requires a great amount of expertise, experience, and trial and error. The first attempt to automatically generate a mechanism assembly that replicates a prescribed motion was for mechanical toys [Zhu et al., 2012]. The user specifies the desired shape, a kinematic chain of its possible motion, and the target motion of specified features. The system then assigns a mechanism from a predefined library according to the type of specified motion, and optimizes its parameters through simulated annealing. In this setting, the target motion and geometry are restricted to be smooth, periodic, and non-colliding. Later, Coros et al. [2013] proposed an interactive framework for computational design of mechanical characters. A functional mechanical character is automatically printed from an input articulated mesh and desired motion curves of the end effectors. A parameterized set of motion assemblies is predefined, and the space of achievable curves is explored in pre-computation. During interactive design, assemblies are retrieved and parameters are optimized to match the desired motion curve, using curve features for comparison. Concurrently, Ceylan et al. [2013] proposed the assembly of automata from motion capture data, where the mechanical elements are mainly off-the-shelf pieces, and not custom printed ones. The motion is approximated through Fourier decomposition of the joint angles of a simple kinematic chain, restricted to three orthogonal planes. Once a mechanism is designed, it turns out it is possible to leverage it to design similarly functioning mechanisms. An interactive framework to retarget a given mechanism of gears was proposed [Zhang et al., 2017]. The system interactively takes user preferences and semantic understanding of the shape into account while automatically ensuring validity, both in terms of integrity and functionality. Functionality is encoded as a set of relationships between mechanism elements, the desired new shape and the actuating environment. The mechanisms are restricted to a predefined set of elements (e.g., gears, bevels, linkages) with continuous parameters.

Instead of exploring the whole range of mechanisms, Thomaszewski et al. [2014] focused only on linkage systems. Such systems can be quite unintuitive since small changes may have significant and surprising effects on the end-effector motion. This work presents an interactive system to design such systems. The input is an explicitly animated skeletal character, with joints and their respective angles for each frame. The system starts by assigning a motor to each joint to reproduce the motion, and tries to minimize the number of motors by adding linkages instead. The user can interactively select from proposed combinatorial options, in order to guide the optimization to be more efficient and the result more aesthetic. A key challenge in such designs is avoiding singularities, or *mechanical locking* of the assembly. It is shown that a lock-free assembly is one that exhibits full-rank Jacobians and therefore Singular Value Decomposition (SVD) is employed to detect and avoid degeneracies during the optimization process. As direct follow-up [Bächer et al., 2015], several high-level operations were added to the system, such as directly designing the motion curves of the end effectors, controlling the overall dimensions of the linkage system etc. Instead of direct simulation, the system is represented only by joint positions and a connectivity graph, where the rest-pose length of neighboring joints must be kept. The system is modeled with an analytic (recursive) expression, which renders deriving and solving it significantly faster, allowing interactive rates.

Walking automata were also investigated. The difficult problem of a stably walking character without sensors was proposed to be handled by using a pre-configured database of mechanisms [Bharaj et al., 2015]. In this method, high-dimensional valid kinematic parameters are learned from the database by using a multivariate Gaussian mixture model (GMM), which guides an expectation-maximization optimization. The optimization considers distance, upright direction, smoothness, effort (force), and similarity to the initial design via Covariance Matrix Adaptation (CMA). Megaro et al. [2015] propose an interactive tool that allows novice users to design a printable walking robot with an arbitrary number of legs, general gait style, and possibly a spine. The robot consists of 3D printed parts and off-the-shelf servo motors. The user can interactively change the dimensions of the limbs, their joints, and the walking style, represented by motion curves and footfall temporal patterns. The system interactively computes a walking sequence that respects all the desired preferences while keeping the object's center of mass within its support polygon for stability. The optimization is done on a kinematic chain representation, common to robotics applications. An illustration of the design system and its results can be seen in Figure 3.6.

Figure 3.6: Megaro et al. [2015] propose an interactive tool for the developments of walking robots. An arbitrary number of legs and gaits are supported. Left: the design of the footfall pattern graph. Middle: preview of the robot's support polygon and center of mass. Right: fabricated prototype. Used with permission.

An interesting aspect of all the publications described here is the representation of motion. Motion has to be conveniently given as input, and efficiently and accurately compared during optimizations. Traditional animation tools such as rigs and linear blend skinning were elegantly used to provide hints to the design systems about possible joint locations. Motion capture is another approach taken for input motion representation. However, like other forms of inputs described here, it was converted to motion curves, which is the dominant representation of choice for intermediate steps, where quality quantification is the main concern. Another interesting point is that all approaches used a prescribed library of mechanisms or joints to control the physical motion. An interesting challenge would be automatic computational development of these libraries.

3.3   HIGH-LEVEL OBJECTIVES

Perhaps the most exciting contribution the computer graphics community can provide in the context of computational fabrication is design through high-level goals. One of the most prominent examples was proposed by Shugrina et al. [2015]. In order to bring the fabrication-aware design process closer to novice users, they propose to control the resulting shape by exposing only a small set of parameters to the designer. A system to explore predefined families of objects is introduced, where professional designers create the designs with many parameters and constraints. In addition, the expert designers define validity tests, as well as a small set of parameters to be exposed to the user. The system adaptively samples the highly multi-dimensional design space (using a k-d tree), and maps the user-exposed parameters to the underlying ones. The object is represented through base geometry generators and a tree of geometry processing operations, which could potentially hold anything from CSG operations to fluid simulations. During the design process, only manufacturable instantiations are allowed, according to the pre-defined validity tests. The user can tune the exposed values, and live feedback provides an indication of further valid exploration ranges, consisting of only valid geometry. In order to speed up the instantiation process, the sampled geometry is cached at sub-tree levels. In follow-up work, this notion was further distilled to avoid jumps that occur during the design exploration process. These originate from the fact that the topology, or connectivity, of the designs is not consistent, and hence jumps are possible when change in parameters translate to different sampling regions. The proposed method smoothly interpolates the geometry of such designs. The smooth interpolation allows for a better exploration experience, and allows for smooth interpolation of pre-simulated physical properties of the samples [Schulz et al., 2017].

In this section, we will explore designing and optimizing shapes under specific non-trivial, non-local objectives. First, we will focus on designing for integral quantities, i.e., physical properties which require integration over the whole object (Section 3.3.1). Then, we will see how the functionality of an object, and the functional relationship between its parts can be used to guide the design process (Section 3.3.2). Last, we will dive into a specific case of functional design, which involves puzzles and other interlocking designs (Section 3.3.3).

3.3.1   INTEGRAL QUANTITIES

Controlling integral quantities, such as total mass (0th moment), center of mass (1st moment), rotational axes and moments of inertia (2nd moment), as well as buoyancy, aerodynamics, and acoustics is a challenging problem, addressed by the publications described in this section. Several different approaches have been taken, addressing the efficient accumulation of properties over the entirety of the designed object. Anything from explicit evaluation per element through regular or adaptive spatial partitioning to elaborate manifold and spherical harmonics are employed in order to tackle these exciting challenges.

Perhaps the first step in this direction was controlling the center of mass of fabricated objects, and thus ensuring their stability when placed on a surface, or hung from a string [Prévost et al., 2013]. Given a triangle mesh and the relative direction of gravity, this method ensures the object's stability by hollowing and deforming the model. Mass and center of mass are computed by integrating over the volume between the outer and inner surfaces of the model. To facilitate carving, a voxel grid represents the inside of the object. To control the center of mass and maintain printability, a heuristic carving scheme is proposed where all voxels outside of a cutting plane are marked as hollow, excluding some of the boundary ones, maintaining minimal wall thickness constraints. User-defined handles are used to deform the surface via linear blend skinning (LBS), and an iterative optimization process alternates between inner voxel carving (discrete) and deformation (continuous). Gradient descent with analytically computed gradients (using triangular meshes with Laplacian coordinates) is used for solving the deformation step and keeping the interface at reasonable frame-rates.

As follow-up work, Bächer et al. [2014] propose to control the principal direction of the moment of inertia, and thus allow arbitrary shapes to spin about a given axis, surprisingly converting a given shape to a spinning top or a yo-yo. Similarly, the optimization process employs carving and deformation. In this method, deformation is performed using an automatically generated cage and the carving optimization is more elaborate, employing sequential linear-quadratic programming (SLQP) over an adaptive and dynamic octree structure. Similarly, linear programming was employed to perform voxel-based hollowing in order to control CoM and center of buoyancy, thus controlling floatation orientation, stability, and height. Musialski et al. [2015] propose controling various integral quantities, such as center of mass, moment of inertia, and buoyancy, by looking at offset surfaces. By computing inward and outward offsets from a given mesh, the different properties can be optimized. The offset directions are primarily orthogonal to the surface, but adjusted to avoid intersections. The inward offset is bounded by the medial axis. For faster and simpler optimization, a dimension-reduction scheme is proposed, employing manifold harmonic decomposition. In follow-up work, a general framework for shape optimization is presented [Musialski et al., 2016a]. The framework supports quantitative target properties, which can be cast to a variety of previously especially solved problems, such as center of mass, center of rotation, buoyancy, and even acoustic frequencies. In addition, the frameworks requires material properties, constraints and a shape parameterization. The latter represents a shape by a set of parameters, determined according to the specific tackled problem, and should help guide the optimization to the correct search directions. For most demonstrated cases, manifold harmonics are used, but other options are also possible, including cage-based or revolution-based parametrizations.

Another interesting aspect that was looked at is aerodynamics. Umetani et al. [2014] introduced an interactive design system to make paper airplanes. Aerodynamics is analyzed locally through "wing elements," and the model is adapted for better flight performance as the user makes changes to the desired shape. Later, Martin et al. [2015] proposed a method for designing kites. Aerodynamics is pre-measured in a data-driven manner, and used to design and analyze new aerodynamic models. Spherical harmonics are employed to extend the aerodynamics model to many directions.

In addition, the design of custom sound filters was explored [Li et al., 2016]. The method is based on a parameterized primitive which is simple and cubic (a cube with six small pipes, one on each face). The acoustic properties of this family of primitives are sampled and computed in a preprocessing step. Given a general 3D shape, an array of these elements is sought such that they follow a prescribed sound filtering behavior. A combinatorial optimization is run on the the interior of the given input shape to compute an array of voxel filters that roughly match the desired target behavior. Then, a continuous optimization fine tunes the parameters of each voxel. The result is a sound filter that dampens or passes different frequencies according to a specific target profile. Recently, a novel approach was proposed which enables the real-time analysis of the resonance frequencies of arbitrary geometry [Allen and Raghuvanshi, 2015]. Employing this approach, a method was proposed for the interactive design of wind instruments [Umetani et al., 2016]. Here, the user provide the internal volume which is used to produce sound, and an interactive system helps in designing finger holes positions and sizes. The end product is a playable wind instrument of arbitrary geometry, producing several tones according to the cavities which are blocked by fingers.

As can be seen, there is no dominant representation for optimization of integral quantities: almost every representation discussed in this book is employed for this end in one way or another. Voxels were employed because they simplify volume and mass computations. Octrees were employed to improve the memory footprint of the latter, at the expense of some computation. Spectral representations were employed to reduce the problem's dimensionality without affecting high-frequency details. Primitives were also used to reduce imposed computations, by restricting the search space. Each method exhibits some benefits, but it seems that there is no clear consensus on any of them yet.

3.3.2 UTILITY

Knowing what an object is used for may be the most important tool to guide a high-level design process. For example, identifying that a door should cover a region when closed can easily guide its dimensions throughout the design process. Similarly, knowing that a part should act as a container allows an automatic system to indicate to the user whether objects placed in it might fall out.

In early research in this context, the valid design space of furniture was considered [Umetani et al., 2012]. An interactive framework was proposed that allows the user to freely design plank-based furniture, while the system provided feedback and suggestions to ensure that the design was stable and durable. The system warns the user when planks are not connected well, the design might topple, or durability is compromised due to excessive force on fasteners. The suggestion system proposes several solutions whenever such violations occur. The method is based on expressing the design space parametrically with constraints. The compact representation allows running physical simulation in real-time, and finding a few meaningful directions to explore when the constraints are violated. Koo et al. [2014] suggest to enhance the design process through high-level relationships between furniture parts, and thus rendering the designed object fabricable. First, a set of relationships (e.g., *cover*, *fit-inside*, etc.) and joints dictating possible relative movement between parts (*slide*, *rotate*, *double pivot*, etc.) are defined. The initial mesh and parts are given by the user, and are restricted to cuboids (since this is aimed for prototyping, which renders this approximation acceptable). During optimization, joint parameters and cuboid sizes are changed to minimize deviation and satisfy constraints, derived from user-given relationships.

Lau et al. [2011] propose to convert a given furniture model to fabricable parts, based on an understanding of their functionality. A grammar for man-made IKEA cabinets and tables is demonstrated. Given an aligned model and its class, the shape is labeled by a set of prescribed types, and a grammar graph that describes the mesh is sought. The labeling process uses voxels, and the graph is used to deduce part roles and relationships between parts to generate the fabrication and assembly instructions. Later, Schulz et al. [2014] further present a paradigm for both designing new fabricable objects from a database of parameterized templates and how to convert an assembly of annotated CAD design to such a database. The database consists of a tree of relationships, a list of connectors, and parameterized parts. The design side includes snapping to probable locations, automatically adding connectors to parts, physical simulation for stability verification, and functionality considerations such as properly closing doors, collision-free motion, and symmetry. In follow-up work, a catalog of parametric parts is used to design novel and arbitrary shaped copters, with varying number of motors, geometries and goals [Du et al., 2016]. The algorithm optimizes the copter geometry (position and orientation of the blades) and controller parameters to ensure validity and stability of flight, while maximizing battery life, payload, cost, and size, according to user request.

Another aspect that was explored is foldability, i.e., whether a functional object can be folded into a specific shape or volume. Li et al. [2015b] have suggested to add hinges to a given segmented mesh, thus making the mesh able to completely flatten out through folding. A more elaborate example is *Boxelization*—a method to transform an object into a cube by folding [Zhou et al., 2014b]. First, the space is voxelized: an initial grid is constructed, followed by a non-uniform voxelization that allows slight deformation to the object to reduce the number of near-empty voxels. Then, the cube parts of the objects are connected in a tree manner, upon which a collision-free path for all voxels to move to a boxed volume is sought. The search is done stochastically (beam search combined with simulated annealing), and template joints are assigned between voxel edges when a folding solution is found. Combining this idea with the previously mentioned notion of high-level relationships, an interactive system to design connected rigidly moving objects was proposed [Garg et al., 2016]. The system detects, warns, and suggests fixes to collisions of the moving objects during the design, alleviating designing an assembly that can transform from one configuration to another (referred to as *reconfigurables*).

Intersections are detected through the use of space-time volumes, sampled at constant temporal intervals.

In addition, a method that automatically produces custom grippers and holders has been proposed [Koyama et al., 2015]. Given a stationary geometry and the object that needs to be held, an automatic part is generated that can be fastened to the stationary geometry and holds the target. Several different types of parameterized template holders and grippers are precomputed, and their strength (how much force and torque is needed to move them) is measured in an example-based, scattered data interpolation manner. An optimization is run to find the gripper with minimum volume that both holds on to the input geometry and provides enough strength. For non-standard geometry, a region growing approach is applied to sufficiently cover the held object. The user can select from different proposals for grippers on both the source and target objects interactively. Additionally, it is possible to leave one axis free for movement, or design a gripper with a cut that can be clicked into place. Some of the possible applications for this method are demonstrated in Figure 3.7.

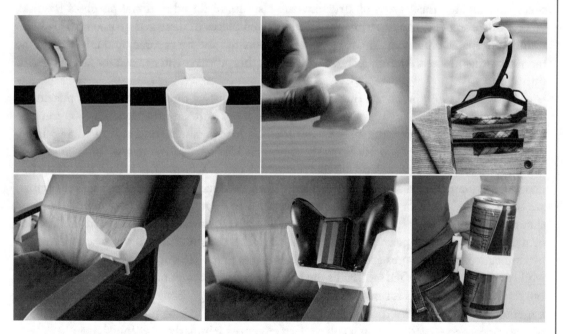

Figure 3.7: Examples of the automatic generation of various grippers and holders, according to the method proposed by Koyama et al. [2015]. From left to right and top to bottom: a mug holder at desk edges, a bunny-shaped coat hanger connected by suction to a window, a game controller holder on a chair arm, and a two-sided holder, connecting to a belt and a soda can. Used with permission.

As seen in the work described above, segmentation, or the indication of parts in the designed object, is imperative for functionality understanding and management. Similarly, defining the right relationships between parts is also a crucial aspect that has been explored in this section. Despite their simplicity, the ideas described here are undoubtedly milestones on the way to pure objective-based optimization. It will be interesting to see whether such functionality descriptions, or other properties interesting to the designer, can be deduced from analyzing large collections of shapes (e.g., *ShapeNet* [Chang et al., 2015]), and exploited as well in this context.

3.3.3   INTERLOCKING ASSEMBLIES

A special case of functionality-based design is a design with interlocking pieces. Such objects have the nice property of not moving, without the need for fasteners. For example, weaving together orthogonal strips of paper or metal sheets can prove to be an efficient approach to fabricating freefrom designs [Takezawa et al., 2016]. Another appealing application for this class is burr puzzles, in which a single *key* piece locks all the pieces together, making assembling them challenging. These types of structures are also useful for educational purposes [Séquin, 2012], where students can experiment with fabricated pieces of dissected geometry. This helps training for spatial understanding, for geometric modeling, and for accuracy and tolerance in the manufacturing process. Alternatively, as we will see, this property can be used for fastening-free furniture.

Lo et al. [2009] propose a framework for creating an interlocking, stable, 3D puzzle out of an input mesh. The puzzle pieces are 2D polyominoes with an offset surface (according to a distance map), creating a thick shell. The mesh is first parameterized and quadrangulated, then a dual graph (or connectivity graph) is constructed for the quads, upon which a polyomino tiling procedure can run in a flood-fill like manner. A construction order is built according to a dependency graph and the shape's medial axis. Tabs and blanks (male and female connectors) are added to the pieces for stability. Burr puzzles can also be automatically computed for a given shape [Xin et al., 2011]. In this method, a template burr mechanism in used to instantiate one or more burr cores in a puzzle. A connectivity graph is constructed and the puzzle is created by connecting the burr cores according to cycles in the graph (in one of four prescribed ways), while ensuring collision-free motion of the pieces. Similarly, computational generation of other interlocking mechanisms for puzzles has been introduced (e.g., Song et al. [2012]) and so has a method to design "twisty" puzzles [Sun and Zheng, 2015] in which pieces can be rotated in different axes about each other in a collision-free manner (similarly to a Rubik's cube).

Recently, Fu et al. [2015] have leveraged such mechanisms to design interlocking furniture, i.e., adding interlocking joints to cuboids-based furniture to create an immobilized structure with only one mobile key part. Given the (only orthogonal) parts of the furniture, they are joined into mid-level groups. Each group is analyzed for immobilization (by an exhaustive search over all pairs), and joints are selected from a library of predefined options. Each group is left with one *key* piece, which is immobilized by a different group. This renders the exhaustive immobilization computation local and therefore feasible, until only one key is left for the whole shape. Note that this type of furniture can be assembled and disassembled many times with minimal wear. Therefore, using the same principles, the work was followed-up, enabling designing multiple pieces of furniture which consist of the same, or have as large set of reused, parts [Song et al., 2017]. Given two meshes representing furniture, they are co-analyzed and deformed for part decomposition which is as similar as possible. Then, the joints are generated through the use of a bipartite graph and identification of small assemblies which can be identical in the different configurations. This enables the large search to be feasible, and produces interesting combinations, such as a ladder, a stool, and a trolley, which consist of the same parts. At the same time, the principles of interlocking mechanisms were used to create fastening-free decorative furniture [Yao et al., 2017]. Given the desired (unconnected) furniture, the user draws on the 2D surfaces designs which are to end up as the interlocking joints. A novel equilibrium analysis is proposed which considers sliding in addition to traditional motion and collapse. This enables allocating the drawn designs to the right parts, and designing the internal joint structure, which produce stable interlocking furniture which still display the designed joints on its surfaces.

One of the grand visions of fabrication-aware design is designing through objectives alone, i.e., having a (novice or expert) user define only what she wishes the object to be or do, with a fully automatic and controllable optimization producing a manufacturable object accordingly. The ideas described throughout this section bring this goal closer. Even though much work still needs to be done for this concept to be realized, the solutions presented here would probably be included as tools in this collective optimization. Below, we discuss problems that arise in different, non-traditional forms of computational fabrication.

3.4 DOMAIN-SPECIFIC APPLICATIONS

While we mostly focus on additive manufacturing technologies, in many cases these can be too expensive (in terms of time or material usage). Here, we explore systems that approximate a given shape, often very coarsely, while exploiting alternative manufacturing technologies. The key goals explored by these systems tend to involve producing visually pleasing approximations, taking advantage of the fact that humans perceive shapes even when only a few of their features are present. A clear example of truly minimalistic shape approximation is that of stable rod structure design [Miguel et al., 2016], in which thick wires are bent to approximate a shape on one hand, but at the same time ensure the sculpture is stable and can be bent and assembled efficiently (Figure 3.8 top left). In this section, we first examine fabrication-oriented shape approximation, followed by two domain-specific cases that are well-studied: planar structures (Section 3.4.1) and architectural construction (Section 3.4.2).

One of the first publications in this context is an interactive tool for designing plush toys [Mori and Igarashi, 2007]. In this system, a prescribed set of simple operations can be

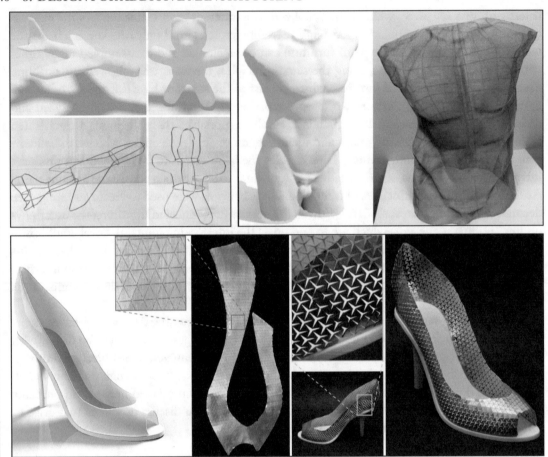

Figure 3.8: Examples of domain-specific design. Top left: bent wire is used to sparsely approximate shapes, while ensuring stability [Miguel et al., 2016]. Top right: designing with wire sheets is surprisingly challenging due to unique restrictions and global effects; this problem is solved through Chebyshev nets [Garg et al., 2014]. Bottom: auxetic materials allow limited stretch due to their special construct, and are addressed through conformal geometry [Konaković et al., 2016]. Used with permission.

performed on a triangle mesh by the user. These operations affect a set of 2D patches, which are generated, deformed, split, or merged, while the 3D mesh is updated as a result of a simplistic simulation run over these patches, between which seam curves define the interfaces. Recently, this notion extended, and a framework to design actuated plush toys was presented [Bern et al., 2017]. Given a 2D plush shape, the user dictates planar motions on the design (although the extension to arbitrary motions should be simple). These are converted to deformations via FEM simulation, and a set of fibers is generated which approximates the desired motion. From this set, co-activation is analyzed and a few different simpler cables-and-motors (*winch-tendon* systems) options are proposed to the user. Similarly, Skouras et al. [2014] present an interactive system to design inflatable (non-stretching) objects. The system automatically finds flat patches to connect, by employing an optimization over FEM simulation, and uses tension field theory to estimate wrinkling effects on simplified versions of the mesh (to speed up optimization and simulation).

Garg et al. [2014] propose a computational approach for designing with wire sheets (woven wires arranged in a regular grid—see Figure 3.8, top right). This type of design is difficult due to several mechanical properties induced by the physical configuration of the material: local changes have global effects, stretch is not possible but shear is, and local shear and curvature are dependent. It turns out that Chebyshev nets present this quality, and a novel representation for discrete Chebyshev nets is introduced. It is further extended to enable design under these constraints, employing several strategies to overcome mathematical and computational difficulties, such as integration, multi-grid, interpolation, etc.

Recently, a method for interactive design via auxetic materials (i.e., flat flexible material that can stretch uniformly up to a certain extent) was presented [Konaković et al., 2016]. This class of materials is realized through cutting non-extensible materials, such as metals or plastic, in a specific regular shape. Elements formed through this cutting scheme can rotate relative to one another, allowing for a limited stretching effect. This enables the approximation of doubly curved surfaces (such as the sphere) using only flat pieces, making it attractive for fabrication. Various possible applications are demonstrated, through an elegant use of conformal geometry, which facilitate otherwise non-trivial or even infeasible optimization (see Figure 3.8, bottom).

3.4.1 PLANAR STRUCTURES

The computer graphics community has also explored the approximation of desired shapes using only planar elements. For physical objects, this traditionally falls in the realm of papercraft [Kilian et al., 2017, Massarwi et al., 2007, Mitani and Suzuki, 2004, Tang et al., 2016]. The emergence of widespread computational fabrication tools such as laser-cutters or Computerized Numerical Control (CNC) devices has given rise to new applications in this field. For example, accurately cutting a single sheet of paper can approximate elaborate shapes with only one fold, generating so-called *pop-up models* [Li et al., 2010, 2011]. Motivated by the need for quick and inexpensive physical prototypes for visualization, planar structures were proposed as well, as portrayed in this section.

Hildebrand et al. [2012] proposed to create such an approximate physical visualization by generating planes based on planar shape cross-sections, sliding them onto one another. This method proposes to iteratively add new planes to the structure until the shape is sufficiently depicted. This approach poses several challenges. In order to approximate the shape well, selected cross-sections are altered to include important nearby geometric features marked by the user.

Feasibility must be ensured at every step, i.e., it must be possible to slide the planes one onto another without obstruction. An extended BSP tree is proposed to facilitate the quick addition of planes into an existing array while ensuring constructibility. Last, insertion order is decided through a branch-and-bound strategy tree. Later, Schwartzburg and Pauly [2013] introduced an interactive design framework for planar structures. By constructing a connectivity graph and running an optimization over plane positions and orientations, constructibility and rigidity are ensured. The user provides an input mesh and is able to modify the constraints (what plane fits with what other plane) interactively during the design process, yielding stable structures for a variety of uses from illustrative figuring to functional furniture without any external support or affixing.

Planar structures can also be built according to cross fields, which are a good local approximation of the shape [Cignoni et al., 2014]. For this method, an input mesh is annotated with a cross field. This field, which can be automatically calculated, indicates the principal directions of the shape throughout the surface. Planar curves are then traced out on the surface, conforming with the cross field while still being as long and as smooth as possible. From these curves, planar *ribbon shape* slices are generated. Stability is estimated through a *divergence* measure, which quantifies orthogonality. Realizability is ensured through the construction of a directed intersection graph, which ensures feasibility when it is acyclic.

McCrae et al. [2014] further present *flatFitFab*, an interaction system for designing planar structures including a static simulation to analyze stress on the slits for fracture avoidance. The design process uses 2D sketching, imposing loads, and procedural modeling, facilitating regularity in examples such as vertebrae. Physical assemblability is verified again through identifying cycles in a graph. An example of fabricated assemblies generated by the system can be seen in Figure 3.9. Of course, planar elements can be used to generate watertight shape approximations and not only cross-sectional ones. Such methods unify similar facing regions of an input mesh to single planes. These can then be held together through internal planar connectors [Chen et al., 2013a] or finger joint connectors [Chen and Sass, 2016].

Planar structures offer a cheap and fast method for producing a plausible representation of a full model. This method is very accessible through the growing popularity of laser cutters. As can be seen in the various results, planar structures have great expressive power for both casual and expert users, and possess a pleasing aesthetic quality. The great advantage of manufacturing only planes unfortunately also imposes the need for a manual assembly step. Thus, ensuring assemblability and stability once the model is built is a major concern for the aforementioned publications, which was commonly addressed by the use of a connectivity graph. One can imagine other applications that might emerge from this line of work, such as using these structures as a skeleton for a shape approximating skin (as mentioned above [Garg et al., 2014]).

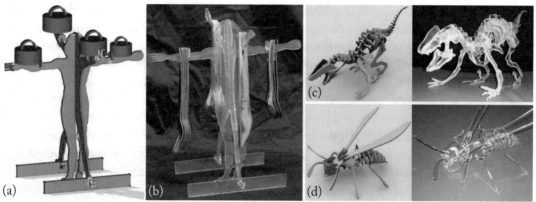

Figure 3.9: Fabricated results generated by the system introduced by McCrae et al. [2014]. The method analyzes slit stress for given loads (a and b), and enables rapid and complex modeling by both experts (c) and amateurs (d). Courtesy of James McCrae.

3.4.2 ARCHITECTURE

As previously mentioned, structural integrity is an uncompromisable aspect of architectural applications. However, design and structural analysis are typically two independent steps, forcing the designer to alternate between them during the development process. The most common tool in architectural design is freeform geometry. This representation ensures smoothness and is convenient for design. Structural analysis is usually done using Finite Element Analysis (FEA), which requires a finite element representation. This is one of the main reasons why interleaving these two steps in uncommon. Interestingly, isogeometric analysis [Cottrell et al., 2009] was developed to bridge the gap between a NURBS representation and FEA. This approach potentially facilitates structural analysis during the design process for CAD applications, but may be just as useful for architectural applications. Regardless, both representations do not consider manufacturing constraints. Most commonly, such constraints are aimed at lowering manufacturing costs through the basic construction elements, or panels. The following work takes this principle into account. Note that architectural design is a wide and extensive field of research, bringing together several research communities. In this part, we will partially cover the publications in the computer graphics community, concentrating on the work most relevant to fabrication-aware design. For a full survey, we refer the reader to existing reports on fabrication-aware design in the architectural context [Pottmann, 2013].

Planar elements, for example, are significantly easier to manufacture than are general non-flat shapes. Liu et al. [2006] explore and expand the use of PQ (Planar Quadrilateral) meshes. This work proposes a method to convert freeform structures to PQ meshes, and further introduces a new sub-class of PQ meshes, called conical meshes, which present properties advantageous to architectural applications. They do this by presenting an optimization framework to perturb the vertices of an input quadrilateral mesh. A set of terms to be optimized is specified, including planarity, fairness, and closeness to the original mesh. Additional terms are also introduced for the conversion to conical meshes, or circular arc meshes (which is another well studied sub-class of PQ meshes). Sequential Quadratic Programming is employed for small meshes (~1 k vertices) and a penalty method solved with a Gauss-Newton method for bigger meshes, due to efficiency considerations. This formulation can also be combined with subdivision surfaces to potentially create a powerful modeling tool. Similarly, Eigensatz et al. [2010] minimize production cost through the approximation of a given freeform surface by easily manufactured panel templates. The method is restricted to planar, cylindrical, paraboloid, toric, and cubic patches, and seeks to minimize the set of required panel types while preserving a user-determined quality level. The solution is an iterative process alternating between continuous (nonlinear least squares via Gauss-Newton) and discrete (set cover problem) optimizations.

Another example for cost reduction in panel manufacturing is repetition of costly parts. A special mold must be made for unusual parts, and if this mold can be re-used then production costs are reduced dramatically. Given a triangular mesh, one can reduce the number of unique template triangles to a given number [Singh and Schaefer, 2010]. This method iteratively finds clusters of triangles (by their representative/canonical polygons) and labels each triangle appropriately. K-means is iteratively employed until the specified number of clusters is found, and the canonical polygons are computed via nonlinear least squares, minimizing the distances between all triangles in the cluster to the canonical one. Finally, the method globally optimizes vertex positions to match canonical polygons by solving a Poisson equation. This is iterated until convergence, and is rather dependent on input triangulation. Similarly, Fu et al. [2010] find a limited set of unique template non-planar quads and template assignments for a given quad surface. They iterate between clustering/generating representative quads and optimizing vertex positions until convergence. Clustering is done by analyzing an edge sharing graph and using a heuristic to solve a minimum K-clustering sum problem. This method employs conjugate gradients for non-linear optimization and is also dependent on the input quadrangulation quality.

Taking a different approach, Pietroni et al. [2014] create statically strong hexagon-dominated grid-shell structures (e.g., for a steel-glass building) based on a Voronoi diagram, weighted by stress. Grid-shells are traditionally strongest for triangle structures, and are commonly designed using quadrilaterals due to aesthetics and other mathematical advantages. This work demonstrates how hexagon-dominated grid-shell structures can be designed to be stronger, while still remaining pleasing. First, a linear static analysis is run, which enables the decomposition of every point to the two principal stress directions (min/max stress, interpreted as a frame-field). This seeds a Voronoi diagram, which is used to compute the resulting grid-shell structure. This process yields structures with excellent static performance, since they are built to align with maximal stress direction. The method further optimizes the structures for cell regularity and symmetry, improving aesthetics.

3.5   UNDERLYING REPRESENTATION ANALYSIS

The work covered by this book employs many different geometric representations to encode shape, materials, motions, hierarchy, abstractions, and annotations. Of course, not all fabrication applications use the same geometric representations. Some work with surfaces, while others work with curves or volumes. Some use regularly sampled representations (e.g., voxels), while others use irregular meshes. Another important aspect is the information attached to the shape, along with the representation. These *attributes* can relate to physical properties of the material (e.g., elasticity or appearance descriptions) or the geometry of the object (e.g., its curvature). The choice of geometric representation and accompanying attributes depend on computational properties required by the application.

In this section, we analyze which geometric representations and attributes are most commonly used for fabrication applications. For each representation, we discuss its computational properties (e.g., what types of computations are most efficient with it) and describe how those properties are leveraged in different fabrication applications. For each attribute, we examine how and why it is or is not used with respect to the applications. In doing so, we hope to provide information to help researchers choose geometric representations for future fabrication projects.

Our analysis focuses on relating computational properties of geometric representations to the requirements of fabrication applications. Theoretically, almost every geometric representation is able to specify almost any shape and perform any geometric operation if given infinite storage and compute time. For example, a voxel grid can represent any 3D function at any resolution with infinitely small voxels. However, in practice, there are significant trade-offs in the accuracy, or *conciseness*, and computational efficiency of different representations. Some require more storage for the same level of accuracy (e.g., voxels usually require more storage than irregular grids). Some representations naturally represent different *ranges of shapes* (e.g., triangle meshes would describe a cube more easily than a sphere). Others enable more efficient computation of *Boolean operations* (e.g., voxels allow immediate volumetric intersection computations), which are important for fabrication-aware design. The ability to compute *differential geometric attributes*, such as curvature, principal directions, or vector fields, is also an important factor in choosing a representation (e.g., algebraic representations provide analytic derivation of them). Similarly, often times a design needs to specify other, *non-geometric attributes* on the object (such as elasticity coefficients), and a representation should support intuitive specification, and interpolation to unspecified regions. Last, another important property in the context of fabrication is the ability to perform *physical simulation*. Figure 3.10c indicates how well each discussed representation performs in these aspects.

In the following parts, we consider the trade-offs in properties provided by different geometric representations and discuss how they affect their suitability for different fabrication applications. Figure 3.10a provides a summary of the ways in which different geometric representations (columns) have been used for different fabrication applications (rows) in work covered in this book. The cells indicate the ratio of surveyed papers that use a given geometric represen-

(a)

Representations / Tasks		Volumes					Surfaces						Primitives		Proc
		Trees	Slices	Skeletons	Grids	Irregular	Facets	Spectral	Dist Func	Algebraic	Subdivision	Height Fields	Catalog	CSG	Procedural
Generalized	Frameworks	0.67			0.67		0.33						0.67	0.67	1.00
Appearance	Light interaction		0.50		0.50				0.50						
	Patterning	0.20			0.20	0.20					0.20				
Deformation and motion	Reinforcement			0.33	0.33	0.67	0.17		0.33						
	Strain analysis			0.08	0.17	0.33	0.08	0.17							
	Deformation behavior			0.08	0.15	0.54	0.31	0.08				0.08	0.23		
	Articulation			0.10	0.10			0.10					0.90		
High level	Integral quantities	0.20	0.10	0.10	0.20	0.30	0.10	0.20					0.40		
	Stable interlocking	0.29		0.14			0.14					0.14	0.29		
	Functionality					0.40							0.80	0.20	
Domain specific	Shape approximation		0.18				0.45			0.27	0.09				
Manufacturing process	(+self) Support		0.20			0.20	0.60		0.20						
	Partitioning, packing	0.25			0.63	0.13			0.38			0.13	0.13		
	Toolpath		0.50		0.50				0.25			0.25			

(b)

	Attributes				
	Parametrization	Differential	Parts +rel	Elasticity	Appearance
Frameworks	0.25				1.00
Light interaction	0.60	0.20			
Patterning					
Reinforcement		0.17		0.33	
Strain analysis				0.67	
Deformation behavior	0.15			0.85	
Articulation					
Integral quantities					
Stable interlocking			0.29		
Functionality				0.80	0.20
Shape approximation	0.18	0.27			
(+self) Support		0.20			
Partitioning, packing			0.13	0.25	
Toolpath					

(c)

	Trees	Slices	Skeletons	Grids	Irregular	Facets	Spectral	Dist Func	Algebraic	Subdivision	Height Fields	Catalog	CSG	Procedural
Conciseness	good	bad	best	bad	good	good	best	bad	best	best	bad	best	best	best
Range of Shapes	par	good	par	par	good	good	good	par	par	par	par	par	bad	bad
Boolean operations	best	par	par	best	par	par	bad	best	good	bad	bad	par	best	par
Non-geometric attributes	good	good	bad	good	good	good	par	good	bad	bad	good	par	par	par
Differential geometric attributes	good	par	bad	good	good	good	par	good	best	good	good	par	par	par
Physical simulation	good	good	par	good	good	par	good	good	par	bad	par	good	bad	bad

Figure 3.10: A breakdown of representation (a) and attribute (b) types that are employed by the covered work, according to the problems they are used to solve. Color intensities and cell values indicate the ratio of solutions that use a specific representation or attribute out of all solutions for the relevant problems. Note that a certain solution might employ more than one representation, and hence the ratios do not sum up to 1.0. Row (and font) size indicates the number of solutions proposed for the same problem set. (c) Performance of each representation with respect to key properties required by computational fabrication related applications.

tation for each fabrication application (i.e., the number of papers that use the representation divided by the total number of papers for the given application). This ratio is both explicitly written in each cell and depicted by its color intensity (redder is higher). Row sizes indicate the quantity of papers discussing a solution to the corresponding problems. Figure 3.10b further illustrates attributes that were assigned or computed over the objects, displayed in the same manner of ratios of solutions to the corresponding problems.

The remainder of this section discusses columns of these tables. We will briefly introduce each representation (Sections 3.5.1–3.5.4) or attribute (Section 3.5.5) type, discuss its properties, and analyze how and why it is used in different fabrication applications.

3.5.1 SOLID

Irregular: Irregular mesh representations offer a great balance between level of detail and accuracy, linearly approximating all regions which are not explicitly specified. For this reason, it is the most common representation for solids in this book. On the other hand, it is less convenient for data interpolation or Boolean operations, which benefit from regularity. This representation includes several flavors. Its simplest and most common form consists of tetrahedra [Bächer et al., 2014, Bickel et al., 2012, Chen et al., 2014, Christiansen et al., 2015, Dumas et al., 2015, Lu et al., 2014, Panetta et al., 2015, Pietroni et al., 2014, Schumacher et al., 2015, Skouras et al., 2013, Stava et al., 2012, Vanek et al., 2014a, Wang et al., 2013, Xu et al., 2015, Zhou et al., 2013]. Tetrahedra, lacking regularity and structure, are almost exclusively employed for physical simulation of general objects, with FEM or simpler, more direct forces-based methods. Cuboids simplify these computations, on the account of expressiveness, and are especially fitting for furniture design, since most parts are cuboid shaped planks anyway [Koo et al., 2014, Umetani et al., 2012]. General polyhedra are only used when an unstructured partitioning of the space into cells occurs, such as one created by a Voronoi decomposition [Martínez et al., 2016, Panozzo et al., 2013].

Grids: Dividing the volume into regular cells is perhaps the most intuitive way to partition a domain. It is widely used in many fields, and is especially common in medical applications such as MRI, X-ray, and ultrasound scans. In this scenario, each cell may contain the density of the scanned material or many scalars, forming a data tensor, for the case of spectroscopy or Doppler scans [Chan et al., 2003, Tiwari et al., 2012]. In the context of fabrication, grids are also a natural choice since in many cases they match the manufacturing technology, in which each voxel in space can independently be void or assigned a specific material. The regularity of this representation facilitates integration, interpolation of attributes, tiling, frequency analysis, derivation, and many other optimization-based applications. For these reasons, it is found in computational fabrication work almost as frequently as irregular meshes. It is used for printer-level descriptions [Brunton et al., 2015, Hergel and Lefebvre, 2014, Hildebrand et al., 2013, Lu et al., 2014, Pereira et al., 2014, Sitthi-Amorn et al., 2015, Vanek et al., 2014a, Yao et al., 2015], multi-grid solving [Dumas et al., 2015, Wu et al., 2015], to facilitate tiling [Bickel et al., 2010, Li et al., 2016, Schumacher et al., 2015], or other high-level optimization goals [Lau et al., 2011, Prévost et al., 2013, Zhou et al., 2014b]. On the other hand, regular grids are not adaptive, and hence demand a lot of memory as the required resolution increases. In addition, they cannot feasibly represent smooth shapes and introduce local artifacts, making them inappropriate for geometric operations and physical simulation.

Trees: Recursive spatial partitioning through tree structures is an efficient way to reduce the burdening memory consumption requirements of grids. This representation suffers from the same disadvantages that grids do, and pays for its conciseness with more involved neighboring queries. In the context of fabrication, Binary Space Partitioning (BSP) trees are naturally used for object partitioning [Hildebrand et al., 2013], but also for assemblability analysis of planar structure designs [Hildebrand et al., 2012]. Octrees are primarily used as multi-resolution versions of 3D grids. Their adaptivity is exploited to reduce memory usage of large homogeneous regions [Bächer et al., 2014]. They are also used [Dumas et al., 2015] to propagate parameterization information across levels for a multi-grid solver. As can be seen, this representation is very sparsely used, in spite of its clear advantage over grids. As fabrication resolution continues to improve, these data structures are likely to be required more frequently, due to increasing memory loads.

Skeletons: Skeletons are thin structures that are typically equidistant to a given shape's boundaries. The skeleton usually emphasizes geometrical and topological properties of the shape [Tagliasacchi et al., 2016]. In the context of this book, two versions of skeletons have been used: rods and beams, and medial axis.

Rods and beams can be represented as a center-curve, and twisting angles. This representation facilitates physical simulation by dramatically reducing dimensionality, and hence is preferable whenever it applies, as can be seen in reported publications [Wang et al., 2013, Zehnder et al., 2016]. Considering the elongated shape's cross-section has been shown to further provide additional flexibility during optimization [Pérez et al., 2015].

The medial axis is a well-studied representation, consisting of a skeleton and offsets from it. This representation does not directly lend itself to most of the operations required in computational fabrication applications, and therefore is not commonly employed. It can, however, be used to reduce dimensions and dictate topology. In some applications, the medial axis is employed to perform geometry processing operations, such as deformation [Chen et al., 2016] and identification of thin parts [Stava et al., 2012]. In others, it is cleanly used to prevent self collisions during object deformation or thickening [Bächer et al., 2012, Lo et al., 2009, Musialski et al., 2015].

Slices: Many 3D printing technologies perform the fabrication process by depositing the material layer by layer. This requires determining the shape's boundary on each layer. For maximum quality, the layer thickness is determined by the manufacturing resolution. These needs give rise to an application specific representation by *slices*. The shape is divided into a, typically regularly distributed, set of parallel planes, aligned with the printing direction. On each slice, the shape is typically represented by closed curves (or *contours*), or a 2D grid. This representation does not offer accuracy, conciseness, or ease of processing, and is usable strictly for guiding the printing process, as can be deduced by its usage [Brunton et al., 2015, Dumas et al., 2014, Hergel and Lefebvre, 2014, Mueller et al., 2014, Reiner et al., 2014, Wang et al., 2015, Zhao et al., 2016a]. For a more detailed review about this topic from a Process Planning perspective, please refer to Section 4.5.

3.5.2 SURFACE

Facets: Facets-based representations, or polygonal meshes, are nearly ubiquitous in computer graphics applications. Most rendering hardware, for example, is designed to handle triangles alone. Therefore, triangle meshes are the form of input to almost all the work mentioned in this book, excluding architectural applications. A significant number of papers, however, does not require the properties of this specific representation, and use triangle meshes simply out of convenience. In our analysis, we tried to distinguish this type of usage from algorithms that explicitly require a facet-based representation, and reported only the latter in Figure 3.10.

Much like their volumetric counterparts, polygonal meshes offer a good balance between level of detail and accuracy. This facilitates most of the discussed computational requirements: linear interpolation of attributes specified on vertices, physical simulation of linear elements, and marking semantic part association and relationships on the elements themselves. For the same reasons, geometric features are more difficult to derive with this representation, however it is so common that many tools have been developed to this end. For these reasons, it is the most common reported surface representation.

This representation also includes several forms, and triangles, much like tetrahedra, are the simplest and most common one [Chen and Sass, 2016, Chen et al., 2016, Deuss et al., 2014, Mori and Igarashi, 2007, Singh and Schaefer, 2010, Skouras et al., 2012, 2014, Vanek et al., 2014b], used for their ease in surface FEM or membrane simulations. Quadrilaterals are preferred for some applications, especially when nearly flat or rectangular, since they match principal directions of the shape and most sampling patterns more naturally. In the context of architectural fabrication, quadrilateral production is often preferred since it facilitates mold reuse [Fu et al., 2010, Liu et al., 2006, Panozzo et al., 2013]. Hexagonal polygons were shown to produce specific mechanical advantages, such as compliance to stretch and shear [Pérez et al., 2015], or extra strength in grid-shell structures [Pietroni et al., 2014]. More elaborate general polygons were used where structure was not a concern [Chen and Sass, 2016, Umetani et al., 2014], e.g., to guide a laser-cutting operation.

Spectral: Spectral mesh processing is a powerful tool for applications such as filtering, shape matching, remeshing, segmentation, parametrization, and many others. For medical purposes, it is employed to more concisely represent dense voxel data, by extracting multi-resolution features [Madabhushi et al., 2005, Tiwari et al., 2012]. In essence, it is an extension of the classical Fourier transform to irregular grids, and typically involves an eigendecomposition of a linear operator over the input mesh [Zhang et al., 2010]. This representation lends itself to detail and resolution reduction, due to direct control over mesh frequencies. On the other hand, it does not allow direct geometry manipulation, including needs such as Boolean operations or physical simulation. Hence, in the context of computational fabrication, it can be used only for specific tasks. All reported usage takes advantage of the unique properties of this representation to substantially reduce computational efforts in non-trivial ways. Modal analysis has been creatively employed to identify weak spots in a design, eliminating the need for explicit FEM simulation, through computing the response over different vibration frequencies [Xu et al., 2015, Zehnder et al., 2016, Zhou et al., 2013]. Manifold harmonics has been successfully employed for dimension reduction in the context of integral quantities optimization, enabling low-frequency modification (allowing for integral control) while leaving high-frequency details intact [Musialski et al., 2015]. In addition, spherical harmonics were employed to extend a data-driven aerodynamics model to unmeasured directions [Martin et al., 2015].

Dist func (distance function): Representing a shape by marking the space with the distance from it has several advantages. For example, Boolean operations are most natural under this scheme, since they can be performed locally and directly on the functions. Perhaps the biggest advantage, which was exploited in the context of this book, is that this representation is agnostic to topology changes. Minimum thickness is a very common requirement in manufacturing processes. Defining the shape as the level-set of a prescribed distance ensures this requirement is fulfilled and allows for direct tessellation regardless of the topology it induces [Lu et al., 2014, Stava et al., 2012]. The same advantage has been exploited to guide a partitioning optimization [Yao et al., 2015]. This representation has also been used to trace out curves along the space, attempting to have them as long as possible, but again with no knowledge of the resulting topology [Pereira et al., 2014, Zhao et al., 2016a]

Algebraic: Algebraic representation is a different form of implicitly describing a surface, by analytically defining relationships between its coordinates or other properties. Most surfaces cannot be represented in such a clean manner, however it is highly desirable to do so if possible. This representation is concise, elegant, and, most importantly, can be analytically derived, which has been leveraged in fabrication applications to significantly improve gradient-based optimization [Eigensatz et al., 2010, Miki et al., 2015]. Special constructs like Chebyshev nets [Garg et al., 2014] and discrete conformal geometry [Konaković et al., 2016] have been elegantly adapted to model specific material behavior, facilitating editing operations otherwise unfeasible. These examples demonstrate how such representations should be sought out and used whenever viable.

Height Fields: Height fields typically represent a surface using a regular planar grid, where each node stores the respective distance of the surface from the plane, in the normal direction. Being regularly sampled, this representation does not offer conciseness. In addition, it cannot directly represent all surfaces, since it stores only one scalar per node. For these reasons, this representation is useful only to facilitate optimization of said height [Bickel et al., 2012, Lo et al., 2009, Sitthi-Amorn et al., 2015, Vanek et al., 2014a].

Subdivision: Subdivision surfaces are smooth free-form surfaces which are generated using recursive rules [Cashman, 2012]. This formalism allows the definition and editing of smooth surfaces in a concise manner through the use of a control mesh. While this is very effective for animation and CAD applications, subdivision surfaces lack utility when physical simulation or geometry processing operations are concerned. For this reason, this representation is unfortunately mostly avoided in the fabrication-aware context, even though it is very common to computer graphics. Liu et al. [2006] have adapted their proposed method to integrate fabrication-aware design within subdivision systems in architectural applications, potentially creating a powerful design tool. Zehnder et al. [2016] have exploited the inherent smoothness of this concise representation to guide the elastic deformation of curves over the represented surface. Recently, basic external calculus operators for subdivision surfaces have been developed [de Goes et al., 2016]. This allows for a wide variety of geometric operations to be performed on the control mesh, significantly reducing computation time for many applications. Hopefully this approach, and others like it, will hasten the integration of fabrication-awareness in subdivision based design.

Parametric: Parametric surfaces, such as NURBS, or other splines [Provatidis, 2005, Sederberg et al., 2003, Wang et al., 2008] offer great ease in modeling and editing smooth surfaces. Similar to subdivision, they are extremely common in CAD and architectural systems, but lack usability for fabrication applications. For this reason, parametric surfaces are seen in this book only as input in architectural application, and are almost immediately tessellated. Recent advances in isogeometric analysis, mentioned in Section 3.4.2, enable physical simulation on NURBS based representation, and might facilitate the use of these representations in a fabrication-aware context.

3.5.3 PRIMITIVE-BASED

CSG: In constructive solid geometry (or CSG), relatively simple primitives are combined by a tree of Boolean and transformation operations that are included directly in the representation. The attractive properties of CSG include conciseness and convenient Boolean properties. Perhaps the biggest advantage of this representation is the powerful control over the shape through high-level parameters, defined on leaves and internal tree nodes. This, along with simple data structures and elegant recursive algorithms makes this representation arguably the most popular one in CAD applications. Unfortunately, this flexible representation is not popular in the computer graphics community since it is cumbersome for any geometric procedure or simulation, and hence is not commonly encountered in this survey. Some publications were able to leverage the parametric nature of this representation, increasing flexibility of pre-defined library primitives [Koyama et al., 2015, Shugrina et al., 2015]. It is possible that this representation will gain popularity also within the field of computer graphics, thanks to its ease of use for designers, as evidenced by generalized frameworks, which do accept it as input [Shugrina et al., 2015, Vidimče et al., 2013].

Catalog: While CSG representations usually consist of simplistic primitives, the use of elaborate pre-defined or pre-computed elements is much more frequent in this book. The classic use-case of predefined elements is designing with already manufactured, off-the-shelf parts [Lau et al., 2011, Schulz et al., 2014]. The more interesting use is constructing a library of parameterized elements, or *database*. This has been repeatedly demonstrated to be effective in constraining an otherwise unfeasible design space, for man and machine alike. It seems that to date there is no automatic way to synthesize articulated motion without manual definitions of joints [Bächer et al., 2012, Bharaj et al., 2015, Calì et al., 2012, Fu et al., 2015, Koo et al., 2014, Koyama et al., 2015, Megaro et al., 2015, Thomaszewski et al., 2014, Zhou et al., 2014b] and other mechanisms such as gears and pulleys [Ceylan et al., 2013, Coros et al., 2013, Zhu et al., 2012], employed for specific pre-assigned cases. Other interdependent mechanisms such as tensegrities and burr puzzle cores also seem to elude automatic generation, and must be pre-configured to allow their design [Gauge et al., 2014, Xin et al., 2011]. Another approach starts by perturbing and measuring the response of fundamental elements, allowing for more elaborate optimization in a later step, which instantiates and modifies them. This effectively constrains the design space, making it feasible for optimization and human manageability [Li et al., 2016, Martin et al., 2015, Panetta et al., 2015, Schumacher et al., 2015].

3.5.4 PROCEDURAL

Procedural: Defining shapes programmatically can be a powerful tool. Potentially, extremely complex structures can be represented by a few lines of code, if they incorporate some reasoning. As manufacturing technologies evolve, designs will probably be spanned over several orders of magnitude, from process resolution through micro-structures to complete designs. Explicitly describing these details is obviously unfeasible, for example, a small cube manufactured by weaved wires even today requires gigabytes of storage to be explicitly represented. Hence, procedural-based representations, where parts of the shape can be generated on-demand, are likely to become popular. To date, however, such representations are not in use, due to the overhead they impose, and the current capability to describe most designs in full. Generalized frameworks [Chen et al., 2013a, Shugrina et al., 2015, Vidimče et al., 2013] offer a programmatic interface to describe shapes, enabling stream-lining information to the manufacturing device, while bounding memory usage.

3.5.5 ATTRIBUTES

Throughout the different solutions reported in this book, different attributes are assigned, propagated, interpolated, and manipulated over the designed objects. The usage of some of these attributes is as expected. For example, **appearance** properties such as BRDFs, translucency profiles or colors, are assigned to objects as targets or measurements, and used in optimization processes strictly for plenoptic applications [Brunton et al., 2015, Papas et al., 2013, Pereira et al., 2014, Reiner et al., 2014]. Similarly, **elasticity** properties, such as Poisson's ratio, Young's modulus, or parameters for other physical models (continuum mechanics, Hart-Smith, etc.), are used solely for elastic physical simulation [Bickel et al., 2010, 2012, Chen et al., 2016, 2014, Christiansen et al., 2015, Dumas et al., 2015, Gauge et al., 2014, Koyama et al., 2015, Lu et al., 2014, Martínez et al., 2015a, Martínez et al., 2016, Panetta et al., 2015, Pérez et al.,

2015, Schumacher et al., 2015, Skouras et al., 2012, 2013, Stava et al., 2012, Wu et al., 2015, Xu et al., 2015]. Interestingly, not all solutions for tasks that seemingly mandate elasticity actually use it. Some strain analysis solutions avoid direct elasticity simulation through modal analysis [Zehnder et al., 2016, Zhou et al., 2013], or geometric properties such as cross-section dimensions [Umetani and Schmidt, 2013]. Even some applications aimed at controlling deformation behavior elude physical simulation by exploiting geometric observations, in order to facilitate real-time response [Mori and Igarashi, 2007, Skouras et al., 2014].

Attributing semantic region assignments, or **parts**, aids in aesthetic shape decomposition, since important features are not disconnected [Vanek et al., 2014a, Xin et al., 2011]. Evidently, it also aids in understanding and specifying object functionality, since semantically similar regions also probably have similar functional goals [Lau et al., 2011]. Further specifying **relationships** between parts (e.g., one part should fit in another, cover another, or support another) has been demonstrated to be very beneficial in the design of functional objects, since it allows for many automatic functionality preserving adjustments to be done during the design process [Koo et al., 2014, Lau et al., 2011, Umetani et al., 2012].

Differential features are also computed and assigned to surfaces. Most geometric attributes, such as curvatures and local shape features, are used to guide semantic shape partitioning or approximation, as can be expected [Hildebrand et al., 2012, Panozzo et al., 2013, Yao et al., 2015]. Vector fields are a well studied tool that provides global properties, such as smoothness or singularities minimization, to functions defined over a surface, while still conforming to local features and principal directions of the shape. This property is nicely leveraged to improve structural stability in the architectural context, where vector fields smoothly interpolate between local stress directions [Panozzo et al., 2013, Pietroni et al., 2014]. They are also elegantly used to trace curves on a surface, being as long and as smooth as possible on one hand, but conform to principal directions (or other directions, prescribed by the user) on the other, in the context of planar structures [Cignoni et al., 2014]. This unique property is likely to be exploited more in the future.

The mapping of a 2D plane to a surface, or **parameterization**, is also employed in various ways. In a classical manner, it is used for texture mapping, in the context of pattern synthesis [Dumas et al., 2015], and for quadrangulation [Lo et al., 2009]. Such a mapping is also used to dictate light routing through the volume [Pereira et al., 2014]. Another use that is unique to fabrication, is for developable patches. Developable patches are important to mimic the properties of materials which can be bent, cut or folded, but not stretched. Manufacturing processes which deform such planar sheets usually involve fabric or metal. For these applications, both the deformed 3D shape and the sheets from which it is created are relevant to the designer, and are typically directly editable. Hence, maintaining a good mapping from one to the other during the design process is an important task, addressed through parametrization several times in this book [Garg et al., 2014, Konaković et al., 2016, Mori and Igarashi, 2007, Skouras et al., 2014].

C H A P T E R   4

Process Planning

We now focus on the various steps of Process Planning, starting with the compliance with fabrication requirements. The subsequent block of PP is the problem of partitioning and packing. Afterward, we review contributions about the choice of part orientation and creation of support structures (both internal and external ones). Slicing methods and their influence on the fabricated parts are then analyzed. Finally, we summarize the key aspects and approaches to the generation of machine instructions.

4.1   MEETING FABRICATION REQUIREMENTS

Herewith we distinguish between requirements of the shape and requirements of its representation. Shape requirements are printer-specific, and define rules for the compatibility of the geometry with the printing hardware. Representation requirements guarantee that the (tessellated) geometry to be printed actually encloses a solid without ambiguity.

4.1.1   SHAPE REQUIREMENTS

Checks   When the input model comes in raster form (e.g., a voxelization), [Telea and Jalba, 2011] provides the means to analyze the shape and identify problematic regions whose size drops below the printing resolution (e.g., thin walls or other tiny features). In a slightly more general setting, [Rolland-Neviere et al., 2013] describes an approach to estimate the thickness of triangulated models.

Thickness in the complementary part of the object is also an issue. If two parts that are supposed to be separated are too close to each other in the digital model, the corresponding printed parts may be fused together during fabrication. That is why thickness in both the object and its complementary space are analyzed in [Cabiddu and Attene, 2017].

Printer-compliant shape adaptation   Besides just *detecting* possible incompatibilities, in Wang and Chen [2013], an algorithm is proposed to actually thicken sheet-like structures so as to make them printable. More recently, a similar approach was proposed in Adhikary and Gurumoorthy [2017] where the STL model is modified by acting on global parameters so as to make all the thin walls thick enough to be fabricated. In Stava et al. [2012], a more comprehensive analysis includes also structural characteristics of the printed prototype: note that this method is not meant to make the model printable, but it shares several aspects with the previous one. In Zhao et al. [2017], a similar stress-based analysis is proposed with a particular focus on shelled objects.

If the model cannot fit into the printing chamber due to its size, Luo et al. [2012] proposes an approach to split the model into parts that can be printed separately and reassembled after printing.

Apparently, no publication deals with the automatic placement of drainage channels for models with internal cavities to be printed with powder bed and other layer solidification technologies.

4.1.2   INPUT REPRESENTATION REQUIREMENTS

Geometry repairing has received increased attention in recent years, not only for 3D printing, but in general for all the scenarios where a "well-behaving" mesh is required (e.g., Finite Element Analysis, advanced shape editing, quad-based remeshing, etc.). Some repairing methods transform the input into an intermediate volumetric representation and construct a new mesh out of it [Bischoff et al., 2005, Chen and Wang, 2013, Ju, 2004]. In a new trend of methods specifically tailored for 3D printing, a 3D mesh is converted into an implicit representation, and all the subsequent operations (including the slicing) are performed on this representation [Huang et al., 2013, Lefebvre and Perchy, 2013]. These methods are very robust but necessarily introduce a distortion. Robustness and precision are indeed major issues in this area, in particular when self-intersections must be removed [Attene, 2014]. In this case some approaches rely on exact arithmetics [Hachenberger et al., 2007], while some others can losslessly convert the input into a finite precision plane-based representation, and then reconstruct a provably good fixed mesh out of it [Campen and Kobbelt, 2010a, Wang and Manocha, 2013a]. When used for 3D printing applications, however, the aforementioned exact approaches are useful only if the input actually encloses a solid, while they are not really suitable to fix meshes with visible open boundaries [Attene, 2010]. For a more comprehensive overview of mesh repairing methods, we point the reader to [Attene et al., 2013] and [Ju, 2009].

Since 3D printing can only produce solid objects, a repairing algorithm must ensure that the resulting mesh actually encloses a solid (Figure 4.1). If the input mesh has open boundaries, a typical solution is to fill the holes in advance and then rely on some of the previously mentioned repairing methods. This approach is employed by one of the most popular web-based mesh fixing services [Microsoft and NetFABB, 2013], but it makes sense only if the boundaries are actually delimiting surface holes: in this case, recent techniques [Jacobson et al., 2013] can properly fill even complex holes with non-simply connected boundaries. Unfortunately, in some cases the designer uses zero-thickness surfaces to represent sheet-like features (e.g., a flag), and for models of this kind a hole-filling approach would produce rather coarse results. These models are better fixed using the algorithm proposed in Attene [(to appear], where the visible part of the input geometry is modified as least as possible, and only where necessary. Another widely used software that performs mesh repairing for 3D printing is Autodesk's Meshmixer [Autodesk,

2011], where the input STL can be successfully fixed even if it has open boundaries but at the cost of an overall approximation due to the global remeshing approach employed.

Figure 4.1: A raw digitized mesh may not enclose a solid due to various defects (left). Mesh repairing algorithms [Attene, 2010] perform little modifications that turn such a raw model to an actual polyhedron that bounds a solid without ambiguity (right). Used with permission.

4.2   PARTITIONING AND PACKING

For AM, in almost all cases material is deposited on top of the partially existing object, or a support structure if nothing should be printed below. In addition, the printing volume is typically quite limited, due to physical constraints. Therefore, partitioning the print object into several parts is an efficient way to reduce print time, support requirements, and accommodate larger prints. Conversely, multiple parts (whether generated by partitioning or existing in the original design) may be packed together into a single printing session, to optimize time efficiency.

4.2.1   PARTITIONING LARGE OBJECTS

When an object is larger than the printing chamber, solutions exist to split it into parts to be printed independently and reassembled afterwards. Luo et al. [2012] present an algorithm for decomposing an object into smaller volumes that fit within the printing volume constraints of a given device. To do so, BSP trees are utilized to find planar cuts via beam search, minimizing several objectives such as number of parts, connector feasibility, aesthetics, and structural soundness. Voxels and tetrahedra are used for structural analysis via FEM, and a distance field inside the object is constructed to check whether connector addition is feasible. The partitioning is done recursively, and stops when all parts fit in the printing volume. An illustration can be seen in Figure 4.2. In similar work [Hao et al., 2011], seams are placed along lines of high curvature so as to minimize their impact on aesthetics.

In both Luo et al. [2012] and Hao et al. [2011], connectors are created to ease the actual reassembly. Conversely, instead of using connectors, glue, or screws, in Song et al. [2015] the subdivision strives to create a part configuration that is assembled via self-interlocking. The parts are firmly connected because of their shape, and may be repeatedly disassembled and reassembled. Song and colleagues have also proposed a method that enables large, efficient prints by connecting printed high-resolution parts and low-resolution interlocking planes, manufactured through faster and cheaper technologies such as laser-cutting [Song et al., 2016].

Figure 4.2: In order to allow for fabrication of objects larger than the printing volume, a partitioning method is proposed to decompose the object into re-assemblable parts, while considering factors including structural soundness and aesthetics [Luo et al., 2012]. Used with permission.

4.2.2 PARTITIONING FOR QUALITY

Even when an object can fit inside a printing volume, it might be advantageous to print it in multiple parts in order to maximize quality. For example, most printing technologies are anisotropic, and there is less resolution and tensile strength in the vertical direction. Thus, different parts of the mesh are best printed at different orientations (as discussed in greater detail in Section 4.3 below). Hildebrand et al. [2013] propose a method for cutting an input mesh along splitting planes to fabricate each part in the best of the three orthogonal orientations. The space is voxelized, and the error is measured on the boundary voxels. An optimization process finds a set of split planes that minimizes the error and number of parts.

Another line of research considers minimizing or eliminating support structures (see Section 4.4), which not only optimizes printing time and cost, but also avoids the tiny defects on the surface left behind after support material is removed. If a shape could be decomposed into *generalized pyramids*, which are flat-based structures with the remaining boundary forming a height function over the base, it could be printed with no support structures. However, the pyramidal decomposition of an arbitrary shape is NP-hard, and often yields too many parts to be a practical solution. As an alternative, Hu and colleagues propose a decomposition into *approximate pyramidal shapes* [Hu et al., 2014]. A uniformly sampled point set is chosen, and each point votes for its best base from a few predetermined directions. Similarly voting regions are merged into *cells*, which in turn are merged into *blocks*, according to an affinity measure. The best decomposition is then sought through the reduction of the problem to the NP-complete *Exact Cover Problem* (ECP), which is solvable in practice. A similar approach is employed in Herholz et al. [2015], where a limited amount of distortion is allowed to minimize the number of parts to be produced.

While the original concept of approximate pyramidal shapes is powerful, it is not appropriate for dealing with *shell* models (e.g., hollowed objects). Hence, in Wei et al. [2017] a skeleton-based algorithm was introduced to partition a 3D shell model into the least number of parts that can be 3D printed without the need for support structures.

4.2.3   PACKING MULTIPLE PARTS

In principle, a 3D model made of multiple parts can be printed all at once, that is, pre-assembled. However, with current technologies this approach might still have several drawbacks:

- the pre-assembled model can be too large to fit the printing chamber, whereas each single constituting part would fit;

- the model is small enough, but it contains tangency parts which would be fused together during the printing process (e.g., this might happen for the rolling elements of bearings);

- the model has none of the aforementioned issues, but printing the assembly would require the insertion of support structures that could not be removed (e.g., because they remain trapped into an inaccessible inner cavity); and

- the model has none of the aforementioned issues, but printing the assembly would require the insertion of support structures that, after removal, would produce rough surfaces (above the tolerances set in the design).

For all these reasons, the current practice for the manufacturing of components made of multiple parts consists of building each part separately, and thus requires reassembly of the physical objects to form the ultimate component. Some elements of the design (e.g., screws, bolts, bearings, etc.) are obtained from external producers that offer a catalogue of standardized components, while all remaining parts are produced by AM. Typically, each part undergoes most of the process planning steps independently. Each so-prepared part (that is, checked, oriented, and supported) is sent to the software that drives the printer and, through this software, an operator places the part in a free portion of the building plate. When the plate is full, or when there are no more parts to print, the operator runs the actual printing process. In this case, the only process planning steps which are common to all the parts are the slicing and the creation of the toolpath. It is also possible to slice one part at a time while using the same layer thickness: this is done, for example, on EOS machines which use the PSW software to place pre-sliced parts on the building platform. Algorithmic approaches exist that try to optimize for both quality and packing efficiency by properly orienting and placing the various parts on the platform [Canellidis et al., 2006, Zhang et al., 2015c].

4.2.4 JOINT PARTITIONING AND PACKING

In order to optimize both printing time and ease of delivering a printed object (e.g., to a customer), it is important to consider the total size of the *pack* containing all printed parts. This requires investigating how a 3D model can be split into easily printable parts that can eventually be tightly packed in a box and reassembled at the destination: the *print, pack, and ship* paradigm.

The algorithm of Attene and colleagues introduces a user-controllable tradeoff between the packing efficiency and the number of parts to be produced [Attene, 2015]. In Vanek et al. [2014a], the focus is mainly on the reduction of the material usage, for which hollowing and orientation optimization are used; however, the eventual packing efficiency is one of the parameters that drives the combinatorial optimization process. Later, another partitioning and packing solution was suggested, which takes printability, high stress regions and mesh features into account [Yao et al., 2015]. In this method, the input mesh is first tetrahedralized and a stress analysis is run on it. Then, the volume is partitioned by a grid, and the stress levels are assigned to the nodes. The mesh is segmented based on geometric features, and a signed distance field is constructed around each segment (on the grid nodes). These fields are used in an optimization to minimize packed height and cut-quality penalty in three iterative individual steps.

Other solutions to joint partitioning and packing have been proposed, based on pyramidal decomposition Chen et al. [2015] and *boxelization* [Zhou et al., 2014b]. In the latter work, a generic shape is converted into a foldable set of nearly cubical parts. The printer can produce the object in its folded configuration, which occupies a smaller volume and thus might fit the printing chamber. Afterward, the printed prototype can be unfolded into the desired shape.

4.3 ORIENTATION

The choice of the building direction is crucial in layered manufacturing as it directly influences the time necessary to print the object, the amount of support structures necessary to sustain the part during the print and the surface quality [Alexander et al., 1998, Pandey et al., 2007, Taufik and Jain, 2013]. The first algorithms to select a proper part orientation date back to the mid 1990s [Allen and Dutta, 1994, Cheng et al., 1995, Frank and Fadel, 1995, Hur and Lee, 1998, Lan et al., 1997, Richard and Crawford, 1995]. Several other methods have been proposed ever since, each one striving to optimize either for one, or a combination of the criteria discussed in Section 2.3.

Directly optimizing for the build direction in the space of all possible orientations is often too complex due to the non smooth nature of the metrics involved [Ezair et al., 2015]. Former approaches used to consider a small number of candidate orientations, either predefined or computed on a shape proxy (e.g., the convex hull). Many recent methods (e.g., Ezair et al. [2015], Morgan et al. [2016], Wang et al. [2016]) share a similar heuristic: they start with a regular sampling of the possible orientations; shortlist the orientations that perform best according to some quality metric, and, starting from each of them, navigate the space of solutions looking for the closest local minimum. The building orientation is eventually defined as the one corresponding to the lowest of the local minima explored by such heuristic.

Table 4.1 summarizes the properties of the orientation optimization techniques we discuss next.

4.3.1   OPTIMIZE FOR COST

Recent works attempt to optimize for the orientation of the part using the volume of the support structures as only metric. Ezair et al. [2015] showed that the resulting function is continuous but non-smooth with respect to the orientation angles. In Ezair et al. [2015] and Khardekar and McMains [2006] two GPU-based volume estimation of supports structures are proposed. Morgan et al. [2016] define the support volume as the sum of the volumes of the prisms generated by extruding the down-facing triangles up to the building plate.

4.3.2   OPTIMIZE FOR FIDELITY

Delfs et al. [2016] propose an orientation system that optimizes for surface roughness, using the mean roughness depth as a proxy to optimize for surface finish. One of the key features of the proposed approach is the ability to prescribe local accuracy requirements, attaching a target surface finish to each triangle in the tessellation. Masood et al. [2000, 2003] proposed two systems that aim to find the building orientation that minimizes the volumetric error (Section 2.3.2).

In Hildebrand et al. [2013], the authors address the problem of finding the proper build orientation for objects to be printed at low resolution via laser cut (cardboard or plywood). Their contribution is the definition of an optimal orthonormal frame that is suited for the decomposition into smaller parts, each of which to be sliced along one of the three directions with small volume loss error.

Other works focus on the artifacts that may be generated when detaching support structures. In Ahn et al. [2007], the authors investigate how to orient the shape so as to minimize post processing (i.e., supports removal and surface finish). Zhang et al. [2015b] introduced a perceptual model to find preferable building directions in 3D printing, so as to place support structures in the least salient parts of the object.

Table 4.1: Techniques for orientation optimization and their properties; see Section 4.3

Method	Optimizes For					
	Cost		Fidelity			Functionality
	Slice Number	Supports Volume	Cusp Height	Volume Loss	Surface Finish	Stress Resiliency
[Delfs et al., 2016]	O	O	O	O	●	O
[Morgan et al., 2016]	O	●	O	O	O	O
[Wang et al., 2016]	O	●	●	O	O	O
[Ezair et al., 2015]	O	●	O	O	O	O
[Ulu et al., 2015]	O	O	O	O	O	●
[Zhang et al., 2015b]	O	O	O	O	●	O
[Umetani and Schmidt, 2013]	O	O	O	O	O	●
[Hildebrand et al., 2013]	O	O	O	●	O	O
[Phatak and Pande, 2012]	●	●	●	O	●	O
[Ahn et al., 2007]	O	O	●	O	●	O
[Byun and Lee, 2006a]	●	●	●	O	O	O
[Byun and Lee, 2006b]	●	●	●	O	O	O
[Thrimurthulu et al., 2004]	●	●	●	O	O	O
[Masood et al., 2003]	O	O	O	●	O	O
[Masood et al., 2000]	O	O	O	●	O	O
[Pham et al., 1999]	●	●	●	O	O	O
[Hur and Lee, 1998]	●	●	●	O	O	O

4.3.3   OPTIMIZE FOR FUNCTIONALITY

Due to the layered nature of the process, the build orientation can significantly affect the performance of the resulting objects, introducing structural anisotropy [Quan et al., 2017]. Ulu et al. [2015] propose a FEM-based building orientation optimization that maximizes the minimum factor of safety (FS) under prescribed loading and boundary conditions. Umetani and Schmidt [2013] propose a cross-sectional structural analysis based on bending momentum equilibrium. This is used in particular for orientation optimization. Their method avoids computationally expensive FEM simulations, and can be plugged into interactive modeling tools to allow users consider structural robustness during incremental trial-and-error design.

4.3.4   OPTIMIZE FOR MIXED FACTORS

Two genetic algorithms to find the optimal part orientation are presented in Phatak and Pande [2012] and Thrimurthulu et al. [2004]. They both formulate the subject of their optimization as a weighted sum of multiple quality criteria, regarding both cost and fidelity. The weights can be finely tuned by the user to tune the importance of each criterion. Similar approaches have been presented in Byun and Lee [2006a,b], however these methods do not scale well with complex mechanical shapes (and non-mechanical shapes) as they consider a very restricted set of candidate orientations computed on the convex hull of the part.

Decision support systems to aid with rapid prototyping choose the best building direction according to their needs are presented in Pham et al. [1999] and Hur and Lee [1998]. Both systems consider multiple criteria that can be prioritized according to the user needs, such as: overhang area, supports volume, build time, and cost. On the negative side, these systems are specialized for CAD shapes and do not scale well on free-form shapes.

4.4   SUPPORT STRUCTURES

Support structures are a key component of process planning. They are used to compensate for some limitations of the manufacturing processes, in particular maximum overhang angles beyond which deposited material falls, and the large increase in time due to printing inner volumes of a part.

In the following we categorize supports into two main categories: disposable *external* supports that assist the fabrication process and are removed afterwards (Section 4.4.1), and *internal* supports that modify the inside of the object to achieve a trade-off between material cost, print time and physical properties (Section 4.4.2).

Note that internal supports are often generated after slicing (Section 4.5) and during toolpath generation (Section 4.6). However, for the sake of clarity we discuss them here, alongside other support structures.

4.4.1   EXTERNAL

External support structures are sacrificial structures that are fabricated alongside the object. After fabrication completes they are chemically or mechanically removed. This usually involves human intervention and therefore is a time-consuming, expensive step. An example of a complex part printed with and without support is shown in Figure 4.3.

Part orientation is a major factor in external support requirements, and therefore is often optimized to reduce the need for supports (see Section 4.3). Researchers have also proposed methods to slightly deform the design so as to reduce [Hu et al., 2015] or to remove [Cacace et al., 2017] support requirements entirely, or to design models that are guaranteed to print without any support [Reiner and Lefebvre, 2016]. However in most cases the process planner has to comply with the input model and some amount of supports remains necessary.

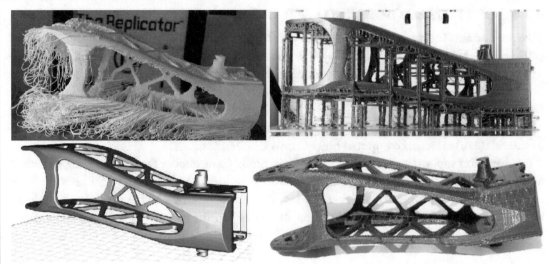

Figure 4.3: Top left: A robot upper leg (3D model below) printed without support. Filament falls due to excessive overhangs. Top right: A sacrificial external support structure, and the cleanup model (below). (3D model from the Poppy project https://www.poppy-project.org/en/, image from Dumas et al. [2014]). Used with permission.

Different types of external supports serve different purposes.

- Local deposition technologies can only deposit material on existing surfaces below. Thus, surfaces appearing mid-air and surfaces at an excessive overhang require support just below them.

- Shapes may move or deform during the fabrication process. This typically happens when fabricated objects are imbalanced and when the raw material (powder, resin) cannot sustain the weight of the print. Another source of distortion are stresses from thermal gradients. To reduce these issues supports acting as fixtures are necessary.

- Some processes can generate a large quantity of heat, in particular metal printing. This excessive heat accumulation results in shape distortions and residual stresses. In such a case, additional supports may act as heat diffusers.

We next review approaches from the literature and discuss their use for each type of support. First, however, let us take a closer look at when and why supports are required.

Islands, overhangs, and self-supporting surfaces. A first situation that requires support is illustrated in Figure 4.4, left. After dividing the shape into layers, one of the slices contains an island—a solid region that appears while not being supported from below. During additive fabrication from bottom to top, the material forming the island will not attach to already so-lidified material. For technologies using material deposition, the material forming the island will fall. As a consequence, material deposited onto the next layer above will also fall, and this will cascade into a catastrophic failure. For technologies using layer solidification, non solidified material below (e.g., powder) will usually be able to support the island—however weight might accumulate over several layers and a heavy disconnected component may start to sink down. For technologies using resins (SLA), the island is problematic: it will typically end up floating in the viscous resin, cascading into a complete print failure.

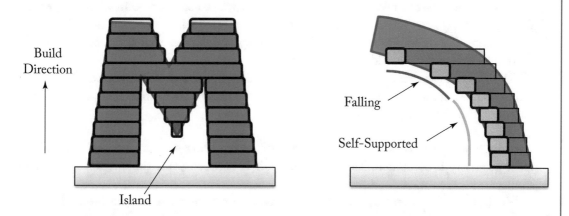

Figure 4.4: Left: The M letter is decomposed into layers. The downward facing tip becomes an island with respect to the build direction. Right: Each layer is solidified from the outside toward the inside. The hatched rectangles are the cross sections of material deposition paths. After a critical angle, the deposited material no longer bonds to the layer below and falls [Livesu et al., 2017]. Used with permission.

Overhangs also can produce problematic cases with material deposition. The material is typically added progressively along deposition paths. The right illustration of Figure 4.4 shows a cross section of the deposited paths for an overhang region. Due to the layering, at some excess overhangs the deposited material will simply fall (red rectangles in the figure). It is interesting to note, however, that until some threshold angle the deposited material will have a sufficient bonding surface with the layer below. This self-supporting property stems from the bonding between successive layers and allows for overhangs to exist up to some maximum angle without requiring support.

Overhangs are less of an issue for technologies employing layer solidification—even though excessive overhangs may distort due to change in material properties during solidification, or to auxiliary motion during preparation of the next layers.

Detecting surfaces requiring support.   Generating supports for areas in overhang requires two main steps: the detection of the surfaces in need for support, and the generation of the support structure itself. For detecting surfaces in overhang a first family of approaches consider the down-facing facets of the input mesh having an angle too steep to print correctly (e.g., Allen and Dutta [1995], Kirschman et al. [1991]). A second family of approaches consists in performing a Boolean difference between two successive slices (e.g., Allison et al. [1988], Chalasani et al. [1995]). The width of the difference determines which regions are self-supporting [Chalasani et al., 1995]. This can also be detected conveniently and efficiently with 2D morphological operations in image space [Chen et al., 2013b, Huang et al., 2014a, 2009a]. As discussed in Huang et al. [2014a], care must be taken, however, with some specific configurations, where some protruding regions might be mis-classified as supported (see Figure 4.5). Dumas et al. [2014] performs the detection directly at the toolpath level, verifying whether each deposited segment is supported by at least half its width from below.

This first analysis generally leads to a compact set of points to be supported. Several approaches then select a subset by down-sampling [Chen et al., 2013b, Dumas et al., 2014, Eggers and Renap, 2007, Huang et al., 2014a].

Generating a support structure.   Once the surfaces requiring support are determined, the support geometry is computed. The main trade-off is between print time, material use, and reliability.

For instance, for filament printers the traditional approach consists in extruding the mesh facets requiring support downward, thus defining a large *support volume*. The support volume is usually printed with a weak infill pattern (see also Section 4.4.2). This still uses a significant amount of material and time, but it is very reliable: the support typically has a large area of contact with both the part and the print bed, ensuring the print stability in most cases. In this context, several approaches modify the support volume to reduce its size. Huang et al. [2009b] use sloped walls instead of straight walls for the sides, shrinking the support volumes in their middle sections. Heide [2010] reduce the support volume by decreasing its size and complexity as the distance below the supported model increases.

Other 3D printing technologies can print complex and thin structures more reliably. In the context of SLA, Eggers and Renap [2007] form a support structure by starting from a regular rhombus mesh filling the print bed. The 3D model is subtracted from the initial structure, removing intersected mesh edges. Points requiring support are attached to the mesh by downward angled beams. Huang et al. [2014a] produce a support made of a sparse set of vertical pillars connected by angled beams for structural strength. The position of the pillars is optimized and their pairwise connections follows a minimal spanning tree (as seen from above) to keep the support structure small.

MeshMixer [Schmidt and Umetani, 2014] builds a thin structure supporting the part in a sparse, limited number of points, generating a support structure resembling a tree. This approach has been used successfully on both FDM and SLA machines. Vanek et al. [2014b] propose an

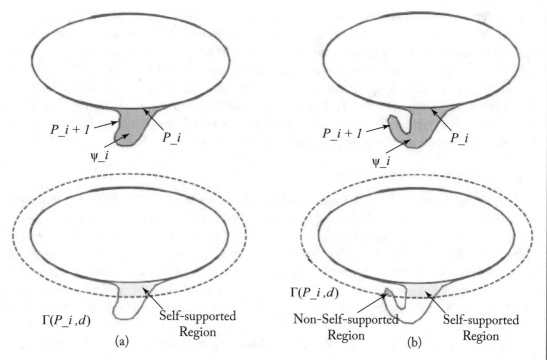

Figure 4.5: The regions in gray are the differences between two successive slices. Regions in yellow are self-supported: there is enough overlap with the slice below during deposition. This may be determined from a dilation of the contour below (a), but care must be taken with features in close proximity (b). Image courtesy of Huang et al. [2014a]. Used with permission.

algorithm to optimize for similar tree support structures. Wang et al. [2013] optimize truss structures for the primary purpose of strengthening 3D printed objects, and extend their approach for support generation. Support beams are added by tracing rays downward. Dumas et al. [2014] generate bridge scaffoldings that rely only on vertical and horizontal bars, improving reliability and part stability while generating small support structures. Remarkably, the horizontal bars can be printed efficiently on FDM printers, but are also well suited for SLA. The static shape balance at all stages of the process is taken into account, enlarging the support structure whenever necessary. Calignano [2014] discusses the design of support structures for SLM (selective laser melting) where supports act both as fixtures and heat sinks.

Support removal. External support structures are meant to be detached from the part after the print is completed. Depending on the materials involved (e.g., metal printing) this operation can be extremely challenging. Furthermore, residual material can remain attached to the surface, badly affecting the surface finish. In polymer AM processes, to alleviate the supports detaching problem, thermoplastic materials that dissolve in alkaline baths are used [Priedeman Jr and Brosch, 2004]. Hildreth et al. [2016] have recently shown that similar approaches are possible even for stainless steel printing, were the differences in the electrochemical stability between different metals can be exploited to dissolve carbon steel supports. Jhabvala et al. [2012] exploited the pulsed laser radiation to print metal supports that are both faster to print and easier to remove; the system supports only SLM printers. A valid alternative consists in trying to orient the part in such a way that the supports necessary to sustain it during the print will stick only to the least salient portions of the shape, as proposed in Zhang et al. [2015b] (Section 4.3.2).

4.4.2   INTERNAL SUPPORTS

The interior of an object is a key factor regarding the material use, print time and mechanical properties of the final result. The impact on material use and print time is easily explained by the fact that the inner volume grows to the cube of the scaling factor (i.e., doubling the size of an object multiplies its volume by eight). Therefore, most of the time and material is spent on the inside of the object. Carving the inside can lead to large savings. However, care must be taken to ensure that the object can still be fabricated (e.g., avoiding the introductions of overhangs or islands, not forming pockets) and that the end result will remain rigid enough. In addition, modifying the structure of the inside of an object gives the opportunity to change its global mechanical behavior, for instance making it flexible or rigid in different places, or changing its balance.

We organize this section as follows. We first consider techniques that focus on creating large empty pockets within objects. Next, we discuss approaches that focus on how to infill the object interior, both with dense and sparse patterns. These techniques have an emphasis on material/time savings and exploit specificities of the processes. Then, we focus on frame structures, which are typically beam or cellular structures optimized to create a strong structure within the object interior. We discuss techniques that focus on changing the mechanical behavior of the object by filling its inside with micro-structures. Since this is a very large topic, we keep the focus on techniques that tightly integrate with the process planning pipeline.

Hollowing.   Most of the techniques we discuss here treat the general geometric problem of computing an inner cavity at a fixed distance from the object surface. The surface of the inner cavity is called the *offset surface* of the model [Farouki, 1985, Rossignac and Requicha, 1986]. These approaches can be used to compute inner cavities that are either left empty or filled with the some infilling pattern, as described in the following paragraphs.

Early approaches obtain a superset of geometric primitives of the offset surface that are trimmed and filtered to form the final offset boundary [Forsyth, 1995]. Qu and Stucker [2003] presented a vertex convolution method for STL files without explicit treatment of self-intersections. Campen and Kobbelt [2010b] introduced an exact convolution approach.

Hollowing can also be performed by computing the *distance field* of the model, and extracting the offset surface from it. Frisken et al. [2000] presented and adaptively sampled distance field. Varadhan and Manocha [2006] approximate the offset surface with a distance field iso-surface extraction, guaranteeing a Hausdorff distance bound on the approximation. Pavić and Kobbelt [2008] traverse an octree distance field and split each cell which is potentially intersected by the offset surface. Liu and Wang [2011] extract the isosurface of a narrow band distance field.

Another class of hollowing methods consider ray representations of solids such as the *dexel structure* [Hook, 1986] or the layered depth normal images (LDNI) [Chen and Wang, 2008]. For a single direction and a uniform grid of rays parallel to that direction, a ray-representation stores the intervals of the rays lying inside the solid. Hui [1994] computes the sweeping of a solid along a trajectory by considering the union of a finite set of ray representations of the solid. Chiu and Tan [1998] computes the morphological erosion of each dexel, taking into account its neighborhood dexels. Hartquist et al. [1999] compute the union of spheres over the boundary of the input model. Wang and Manocha [2013b] place spheres on the samples of a LDNI structure, and compute their union efficiently on the GPU. Chen and Wang [2011] generate a convolution geometric primitives of the offset surface, constructs their LDNI, and filters the points of the superset that belong to the offset surface. Martínez et al. [2015b] consider the dilation of a dexel structure along different successive ray directions, and exploit the winding number of specially constructed meshes. Other methods generate a voxelization of the offset surface. Li and McMains [2014] present a GPU approach to compute the Minkowski sum of a polyhedra by computing pairwise Minkowski sums, and obtain a voxelization of their union. Hollowing can also be done by considering the offset of surface points. Lien [2008] computes the Minkowski sum between two surfaces sampled by points, and distinguishes the interior and boundary points. Calderon and Boubekeur [2014] introduced a set of morphological operations for point clouds (see Figure 4.6).

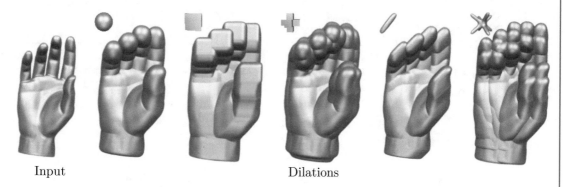

Input   Dilations

Figure 4.6: Morphological dilation of an input point cloud, considering different structuring elements. From Calderon and Boubekeur [2014]. Image courtesy of Tamy Boubekeur.

A few methods hollow the model during slicing, that is at slice level. McMains et al. [2000] consider regularized Boolean operations of each slice contour, in order to approximate the offset surface. In order to achieve uniform hollowing thickness, Park [2005] considers the erosion of a circle swept over the slice contours.

We focused here on hollowing at the process planning stage. As mentioned in Section 2.3.3 hollowing can also be used at design time to change the mass distribution of an object and optimize for various properties such as balance. The techniques discussed here also do not take fabrication constraints into account, such as self-supportability. We discuss in the paragraph *Sparse infill*, hereafter, methods that consider these aspects.

Dense infills. Most technologies such as SLS, SLA, binder deposition and fused filament deposition support dense infilling. On raster devices, it suffices to produce an image of the filled layer contour. On vector devices, densely filling the object requires to follow a space filling curve when solidifying the material. The curve represents the path followed by the deposition device, while it deposits (or solidifies) a wide and thick track of material. The thickness matches the layer height, while the width depends on the technology (typical values are 20 μm for focused lasers, 400 μm for plastic extrusion). Thus, the spacing between two neighboring paths has to match the deposition width. A smaller spacing produces excess deposition/solidification (*overflow*) while a larger spacing leaves gaps (*underflow*).

The shape of the space filling curve can have an impact on both the print time and the final object strength. Some possible patterns are illustrated in Figure 4.7.

Direction Parallel   Contour Parallel   Hilbert Curve   Fermat Spiral

Figure 4.7: Four different filling patterns used in AM. From left to right: direction parallel, contour parallel, the Hilbert space filling curve, and the recently proposed Fermat spiral [Livesu et al., 2017]. Used with permission.

For most technologies—including fused filament deposition—a popular pattern is the so-called *direction parallel* (or *zig-zag*) toolpath [McMains et al., 2000], which consists in filling the slice area with a set of equally spaced segments parallel to one another and linked at one of their extremities. The spacing between the hatches is adjustable to control the final density, and the direction of the hatches changes every two slices to avoid strong mechanical biases. To gain speed or save material, the parameters used during infilling may be different, e.g., reducing binder flow, or changing the laser focus to solidify larger tracks.

Contour parallel infill [Held et al., 1994, Kao and Prinz, 1998, Yang et al., 2002, Zou et al., 2014] is a common alternative to direction parallel infill. The contour parallel pattern consists in a set of concentric closed curves that emanate from the outer boundary of the slice and propagate inwards (see Figure 4.7). These are in fact the offset contours from the layer boundary. This is an appealing pattern since it closely follows the outer contour of the slices. Unfortunately, it also tends to leave gaps within the slices. Indeed, the offset curves from the outer contour meet around the medial axis, and it is unlikely that the remaining space matches the deposition width. These gaps are difficult to handle. For this reason most slicers rely on a hybrid approach that combines direction parallel and contour parallel [McMains et al., 2000]. A few contour parallel curves are produced near the boundary of the slice, and the remaining inner polygon is filled the interior with a zig-zag pattern. The most critical part in the implementation of such hybrid approaches is the handling of the meeting point between the contour parallel and the direction parallel toolpaths, in which detaching, under and overfilling may occur [Jin et al., 2013].

Infill patterns can lead to two issues. First, they may result in many stops and restarts of material deposition, which can lead to several problems (e.g., many small gaps, material feeder failure). Second, sharp turns may slow down the motion or introduce vibrations. For instance, direction parallel toolpaths can lead to many small segments if the slice contains thin walls orthogonal to the deposition direction—inducing a large number of turns. The speed of different dense infill strategies have been compared, including variants that smooth out the sharp turns between hatches [El-Midany et al., 2006, Kim and Choi, 2002]. To obtain a more continuous pattern, Ding et al. [2014] decompose the slice in different regions that use different directions of parallel infill. The patterns from one region to the next are smoothly connected. Zhao et al. [2016a] advocate for the use of spirals. A key advantage is to reduce the number of sharp turns thereby enabling faster motions, while achieving a fill pattern that resembles the contour parallel infill.

Other alternatives to these dominant patterns are Hilbert spiral [Griffiths, 1994], Moore's spirals [Cox et al., 1994], and Peano spirals [Makhanov and Anotaipaiboon, 2007]. A common drawback for both spirals and contour parallel patterns is the lack of *direction bias*, which prevents the generation of the cross weaved layouts used by direction parallel approaches.

Sparse infills.   Since the cost of AM is essentially driven by time and material use, sparse infilling is an important feature. The geometry of a sparse infill is often determined by the target process. In particular, powder and resin systems (e.g., SLS/SLA) cannot create closed voids: non-solidified material is trapped and cannot exit, the sparsity would be lost. However, they can print geometries that are significantly more complex than, e.g., fused filament fabrication. This has led researchers to propose internal frame structures as well as micro—structures, that we both describe in the following paragraphs.

Fused filament fabrication (FFF) and similar technologies (e.g., contour crafting) require dedicated infill patterns, due to the strong overhang constraints. In addition, the thin slanted beams that are used with other processes are slow and less reliable to print. Most slicing software for FFF supports sparse infills. These infills are usually 2D zig-zag hatching patterns that are vertically extruded [Dinh et al., 2015], again possibly changing orientation every two slices to avoid strong mechanical biases. Kumar et al. [2009] explores hierarchical versions of space filling curves to grade the density of the infill pattern.

These patterns are efficient and simple to compute during slicing, and offer a good support for the roofs of the inner cavity. However, due to the vertical extrusion they are not mechanically strong if pressured on the sides. To obtain stronger patterns, Steuben et al. [2016] define the infill as the iso-contours of a scalar field, or the principal directions of a vector field within each slice. For instance, the infill paths can follow the principal directions of stress from a finite element simulation, resulting in stronger patterns for a given load scenario.

There has also been several attempts to move beyond simple vertical extrusions. 3D printing enthusiasts have experimented with interesting 3D infill patterns [Nicoll, 2011]. The software *Slic3r* proposes an infill pattern that produces a 3D honeycomb pattern. This recently led to a new type of pattern called *rhombic infill*. These are formed by the intersection of at least three sets of parallel planes in space. These infill have a number of interesting properties. First, they can be efficiently generated during the slicing process, and by carefully choosing the angles of the inner planes they can be printed as fast as vertical extrusions [Lefebvre, 2015]. Second, when printed with uniform density they are very strong thanks to the inner 3D cell structures. Third, they can be subdivided to locally increase the infill density, as shown in Figure 4.8, right. Wu et al. [2016b] propose a criterion based on the overall rigidity and balance of the object to perform this subdivision. Lee and Lee [2016] and Lee et al. [2017] subdivide closer to the cavity roofs in order to create large empty cavities. They further reduce the size of the structure by removing faces that are not required to support a structure above.

In order to make infill as sparse as possible, Hornus et al. [2016] propose a method that creates maximal inner carvings while ensuring that they remain fabricable with filament deposition (the cavities are self-supporting). This is achieved through morphological operations on the slices, growing a self-supporting inner cavity from the object tops. They further reduce material use by iterating the method within remaining cavities [Hornus and Lefebvre, 2017]. The method tightly integrates with the slicer. Wang et al. [2017] optimize similar nested, self-supporting cavities for various properties such as shape balance. Several other methods optimize for self-supporting inner cavities while considering functional objectives [Xie and Chen, 2017, Yang et al., 2017].

Internal frame structures.   As mentioned earlier, frame structures are especially well suited for SLA/SLS technologies, even though they have also been successfully demonstrated on fused filament fabrication. Wang et al. [2013] propose to fill an object interior with a sparse truss structure. The structure is optimized to reduce the number of beams while preserving rigidity. Zhang et al. [2015a] similarly produce an inner structure made of beams but instead exploit the medial axis of the object, using it as a backbone structure. Medeiros e Sá et al. [2015] generate

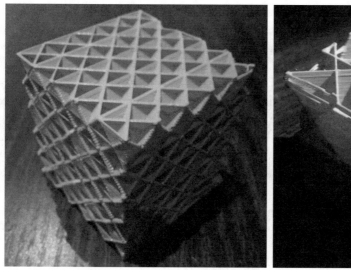

Figure 4.8: Rhombic infill and its hierarchical version [Lefebvre, 2015]. Used with permission.

an adaptive tessellation of the interior, and offset the edges of either the primal or the dual to produce an inner beam structure. Thanks to the adaptive tessellation, the structure is denser along the shape boundary than on the inside. Lu et al. [2014] optimize for a Voronoi diagram inside the print, whose faces form an infill pattern.

A difficulty in designing 3D infill patterns is that their complexity might lead to increased print time, for the same density (this is less true on systems such as SLA/DLP where the entire layer is exposed at once). Yaman et al. [2016] consider how to print efficiently the faces of a Voronoi diagram, following an Euler cycle to solidify in sequence the segments forming the faces in each slice.

Microstructures. Microstructures are internal infill patterns that seek to change the macroscopic physical behavior of the final object. For instance, even when printing with a single material certain microstructures modify the elastic behavior of the object, making it more or less flexible. It is often possible to grade and control the change across the final object.

The design of microstructures with tailored properties was introduced in the 1990s [Cadman et al., 2013, Sigmund, 1995]. A large range of techniques deal with optimizing functional microstructures, such as functionally graded materials for CAD applications [Jackson et al., 1999, Kou and Tan, 2007, Oxman, 2011] and porous scaffold design for bioengineering [Hollister, 2005], among many other applications. A complete review of the field falls out of the scope of this document. Instead, we focus here only on the techniques working in conjunction with the process planning—that is methods within the scope of design for AM [Rosen, 2007].

In particular, as the size of microstructures becomes smaller, approaches explicitly storing the microstructure geometry become computationally infeasible.

Chen [2007] defines micro-structures as periodic tiles that are then efficiently mapped into the volume, similarly to volume texture mapping. By deforming the mapping, the infill locally adapts to a density field. Pasko et al. [2011] considered procedural definitions of periodic microstructures. The parameters of the microstructures can vary spatially to produce graded materials [Fryazinov et al., 2013], for instance to reinforce an object following a cross-sectional stress analysis [Li et al., 2015a]. OpenFAB [Vidimče et al., 2013] provides a specialized language to describe procedural microstructures. The geometric details are efficiently evaluated at slicing time, streaming voxels to the printer. The advantages of procedural representations for the process planning have been identified early by Park et al. [2001] in the context of multi-material fabrication (see also Section 4.5.2).

Different works seek to produce microstructures that can be fabricated [Andreassen et al., 2014, Zhou and Li, 2008] and produce a prescribed elasticity. Schumacher et al. [2015] and Panetta et al. [2015] consider periodic tilings of precomputed microstructures that cover a large spectrum of elastic behaviors (see Figure 4.9). Martínez et al. [2016] consider procedural Voronoi-based microstructures, that can be fabricated with SLA/SLS. They generalize this approach to foams with orthotropic elastic behavior, that can be freely oriented and graded in space [Martínez et al., 2017]. Zhu et al. [2017] explore how to incorporate multi-material micro-structures within a coarse scale topology optimization, taking into account the material space spawned by the micro-structures. Wu et al. [2017] perform high-resolution topology optimization while constraining the average local density, thereby obtaining porous structures resembling bone structures.

Micro-structures are a very promising field of research, with many potential applications [Raney and Lewis, 2015]. We envision that efficient techniques for designing and fabricating micro-structures will tightly integrate with process planning.

Figure 4.9: Printed microstructures with a precomputed elastic behavior. From Schumacher et al. [2015]. Used with permission.

4.5   SLICING

Slicing is central to the process planning pipeline, as it is the step where the 3D geometry is divided into a set of planar contours. These contours will be later manufactured by material deposition. In the following, we assume that the build direction is along the z axis, aligned with the height of the object. Each slice is a plane intersecting the shape at a given height. Assuming the shape is a solid, then its intersection with a slice plane is a closed 2D contour.

There are two important questions to solve when considering slicing: how to determine the set of slices to use and their vertical position, and how to efficiently compute the contour within each slice given the input shape and the set of slices. The two following sections discuss each of these steps.

4.5.1   UNIFORM AND ADAPTIVE SLICING

The simplest approach to divide the object into slices is to subdivide it uniformly, as illustrated in Figure 4.10, left. Given a manufacturing layer thickness τ and an object height H, the object is divided into $N = \lceil \frac{H}{\tau} \rceil$ slices. Each slice i is then located at height $z_i = \frac{i+0.5}{N}$, which is the position of the plane that will be intersected with the object. This approach is widely adopted and most softwares offer it as a standard approach.

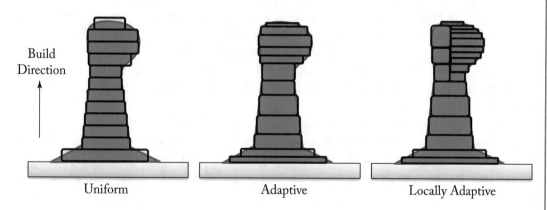

Figure 4.10: A same object sliced with different approaches. Left: uniform slicing (12 slices). Middle: adaptive slicing (12 slices, volume error is reduced). Right: locally adaptive slicing, where the top region is split into two sub-regions sliced independently [Livesu et al., 2017]. Used with permission.

However, many objects have a shape that varies greatly along its height. Therefore, in some regions uniform slicing might use too many slices, while it does not properly capture the shape in others. In particular, surfaces that are slanted with respect to the build direction produce a staircase defect through AM. These areas require to use very thin slices (small τ). Uniform slicing forces the same small value of τ to be used throughout the entire part, thus producing a large number of slices and increasing manufacturing time.

Most technologies are able to change the layer height during manufacturing. *Adaptive slicing* approaches exploit this property, by adapting the thickness of each slice to the shape geometry, as illustrated in Figure 4.10, middle. Given a model for the geometric error (see Section 2.3.2), these approaches refine or coarsen the slices to meet a quality constraint while reducing print time. This can be achieved by locally determining the slice thickness from the error [Dolenc and Mäkelä, 1994, Suh et al., 1994], by subdividing the slices from the coarsest uniform set of slices [Hope et al., 1997, Kulkarni and Dutta, 1996, Sabourin et al., 1996], by merging slices starting from the thinnest uniform slices [Hayasi and Asiabanpour, 2013], or by formulating a global optimization problem [Sikder et al., 2015, Wang et al., 2015]. In a discrete setting the problem can be optimized exactly by dynamic programming, minimizing an error over all possible choices of slices [Alexa et al., 2017].

While adaptive slicing is able to adapt to shape changes along the build direction, it still cannot adapt to a change in part complexity *within* the layer. Consider an object with a vertical wall on the left, and a slanted surface on the right. The vertical wall could be printed with thick slices, however the slope on the right imposes the use of thin slices to limit the staircase effect. To reduce this issue, *locally adaptive slicing* has been proposed by Tyberg and Bøhn [1998]. The key idea is to first subdivide the object into different regions, each region being sliced independently, as illustrated in Figure 4.10, right. Depending on the target technology, the different regions can be built together by locally changing the layer height [Tyberg and Bøhn, 1998, Wang et al., 2015], or they can be printed independently and later assembled [Hildebrand et al., 2013, Wang et al., 2016]. The main issue with this approach is that seams appear along the surface where different layer thicknesses meet. For techniques where parts are printed separately a manual assembly step is required.

Interestingly, a similar approach was used on the object interior by Sabourin et al. [1997]: since the inside is never visible, it can be sliced using a larger thickness than the exterior. This idea can be combined with the aforementioned techniques, for instance performing local adaptive slicing only on the exterior shell while the interior uses the maximal thickness [Mani et al., 1999].

4.5.2   SLICE CONTOURING

Once the set of slices is determined, each slice plane has to be intersected with the input geometry. This operation strongly depends on the representation of the input. We first discuss slicing of triangle meshes and ray-representations (ray-reps). Both are well studied and successful approaches, which are often described as *indirect* approaches: the input CAD model has to be converted into a triangle mesh (tessellated) or a ray-rep (ray-tracing or rasterization). Both conversions require the user to set a precision parameter and may loose information. We discuss these issues in more details below. Therefore, a number of *direct* slicing techniques have been proposed, that avoid any re-sampling of the initial CAD model. We discuss these contouring techniques last.

Contouring triangle meshes. A general scheme for contouring triangle meshes consists in first extracting all intersection segments between the slice plane and the triangles, and then forming loops [Kirschman and Jara-Almonte, 1992]. If the input correctly defines a non self-intersecting solid, the loops will be closed and non-intersecting. Otherwise, a mesh repair step is required (see Section 4.1.2). Alternatively, the slicer may attempt to close holes between nearby segments and resolve intersections.

Implementations mainly differ by how segments and loops are formed. Kirschman and Jara-Almonte [1992] propose a parallel implementation that intersects each slice plane with all triangles. McMains and Séquin [1999] propose an efficient algorithm that exploits the mesh connectivity to sweep a slicing plane through the triangles. The observation is that the topology of the 2D contours remains the same between vertices of the input mesh. Therefore, the contours can be very efficiently produced for all slicing planes in between two vertices. The update to be performed at each mesh vertex is often limited and fast. The open source software *CuraEngine* traverses triangles first, and each is intersected by the slicing planes it covers (see `Slicer::Slicer` and `project2D` in `slicer.cpp/.h`). The segments are identified by the faces to which they belong, and contours are formed by looping over segments following mesh connectivity (see `makeBasicPolygonLoop` in `slicer.cpp`). Zhang and Joshi [2015] argue that mesh connectivity might be expensive to obtain and propose to incrementally construct the contours while triangles are traversed. Linked lists of segments are augmented by adding the new segment to the head or tail (or starting a new list). The segments are not identified directly by intersection points, but instead by the triangle edge to which they belong—making the approach robust to numerical errors in intersection computations.

Contouring of ray-representations. Ray-reps are techniques where the geometry is captured by solid/empty intervals along a set of rays. This technique was pioneered by Hook [1986] who proposed the *dexel-buffer*. This data-structure is built by intersecting axis-aligned rays along one direction with the geometry. Given a closed geometry, the number of intersections is even and each interval can be classified as inside (solid) or outside (empty). The dexel-buffer is closely related to the A-buffer [Carpenter, 1984]. There are therefore interesting similarities between ray-reps and A-buffer techniques for order independent transparency [Maule et al., 2011]. For instance, dexel-buffers can be constructed by rasterization, recording all fragments drawn in every pixel [Lefebvre et al., 2014]. Other efficient approaches are based on Layered depth images (LDI) [Shade et al., 1998], which capture the geometry as a set of depth images each storing a surface sheet in space (a *depth peel*). This data structure has been extended to solid modeling for AM by Chen and Wang [2008], adding normal information along with depth information (layer depth *normal* images, or LDNI). Fast construction methods have been proposed, e.g., using single-pass voxelization [Leung and Wang, 2013] and compaction for increase memory efficiency [Zhao et al., 2011]. Most ray-reps used multiple directions to better reproduce geometries [Benouamer and Michelucci, 1997].

Once obtained, ray-reps may be directly rendered [Hook, 1986, Lefebvre, 2013, Wang et al., 2010] or converted into meshes through efficient procedures [Wang et al., 2010, Wang, 2011, Zhang and Leu, 2009]. However, they may also be directly contoured to extract slices. Different strategies have been used: single set of rays from the object side [Zeng et al., 2011, Zhu and Yu, 2001], rays from two sides [Qi et al., 2013], and rays from the object bottom [Lefebvre, 2013]. Contouring is then performed by marching along the rays, forming polygonal loops [Yuksek et al., 2008, Zhang et al., 2007]. A key issue when using ray-reps for contouring is to decide upon the resolution required to properly capture the geometry and its topology [Huang et al., 2013] (see Figure 4.11). The contours extracted from ray-reps have typically many small segments—each slice is an image and contours are extracted as the outline of the solid pixels. Huang et al. [2013] describe a topology preserving contour simplification, and a full image based pipeline for AM [Huang et al., 2014a]. Another difficulty is the large memory requirements, which is roughly proportional to the surface area. To avoid saturating memory, tiling schemes may be used [Chen and Wang, 2013].

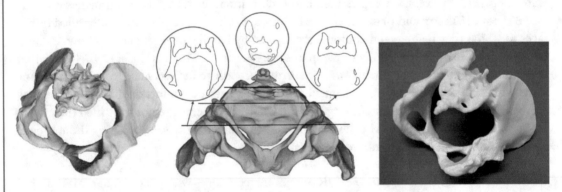

Figure 4.11: Slicing and contouring of a ray representation LDNI while preserving its topology. From Huang et al. [2013]. Used with permission.

Ray-reps have other significant advantages for AM, for instance to perform CSG between complex geometries [Benouamer and Michelucci, 1997, Lefebvre, 2013, Wang et al., 2010], for regulating solids having self-intersections [Chen and Wang, 2013] or for computing offset surface for hollowing parts (detailed in Section 4.4.2).

Direct slicing.   To avoid having to re-sample the CAD model into a triangle mesh or ray-rep, several techniques have been proposed to extract contours directly from the initial geometry. These approaches directly output slice data to the printers. The slice file format is often vendor-dependent, but some independent formats such as CLI (Common Layer Interface) can be used. Open-source printers often accept slice data (e.g., G-Code for filament printers or images for

DLP printers such as the B9Creator or Autodesk Ember). Jamieson and Hacker [1995] provide an in-depth discussion of the pros and cons of direct slicing of CAD models for different input types.

NURBS surface are a common surface representation in CAD software. Therefore, approaches have been proposed for the direct slicing of NURBS models in order to avoid a global tessellation of the geometry [Vuyyuru et al., 1992]. Later approaches consider specialized adaptive slicing [Ma et al., 2004] and orientation procedures [Starly et al., 2005].

Techniques have also been proposed for point clouds, which are often obtained from 3D scanners or vision algorithm. They are challenging to print since the connectivity and topology of the surface is unknown. Early methods project points around the 2D slicing plane and reconstruct a contour [Liu et al., 2003, Shin et al., 2004, Wu et al., 2004]. Yang and Qian [2008] propose to rely on the moving least square method to implicitly define the surface from the point cloud. Qiu et al. [2011] and Yang et al. [2010] refine this approach by considering the global topology of the shape to detect and capture extremal points in an adaptive slicing strategy (see Figure 4.12). Chen et al. [2013b] propose a complete system for scanning an object and fabricating it in a different location, based on point clouds. This includes a novel support generation algorithm.

Finally, Rosen [2007] proposed to apply direct slicing on microstructure lattices, thereby bypassing the mesh generation step. The lattice beams are directly interesected by the slice plane, and thus only the lattice graph is stored in memory. This makes slicing much more efficient and accurate on such objects.

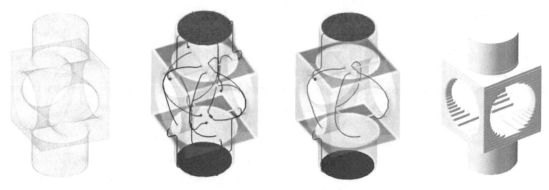

Figure 4.12: Direct slicing (right) of a point cloud (left) via topological analysis (middle). From [Yang et al., 2010]. Used with permission.

Non-planar approaches. There has been a number of interesting attempts at moving beyond standard layers for additive fabrication. These attempts usually focus on a specific technology since they exploit properties such as the ability to perform Z-motion during deposition (fused filament fabrication) or the ability to partially cure material (SLA/DLP).

In the context of SLA, Pan et al. [2012] and Pan and Chen [2015] exploit the formation of a meniscus when an object moves out of the resin tank. The meniscus is cured to fill the creases between two layers. Repeating this process produces smooth, accurate surfaces. Park et al. [2011] show how dithering can cure resin partially and produce slanted surfaces along a layer.

In the context of filament deposition Chakraborty et al. [2008] proposed curved layer deposition with the objective of strengthening shell-like parts by aligning the toolpaths with the surface. The mechanical properties of the parts are discussed in Singamneni et al. [2012]. An interesting question is then to combine flat and curved layers, as discussed in Allen and Trask [2015], Huang and Singamneni [2012]. A key challenge of curved layers deposition is that the curvature of the paths is large as they flow along the surface several millimetres up and down. This has been demonstrated for specific parts, however in a general setting this degree of freedom is challenging to exploit. First, because a novel type of slicers has to be developed, and second because the current design of deposition nozzle complicates the task: collisions between already printed paths and currently deposited paths become possible [Allen and Trask, 2015, Chakraborty et al., 2008].

Beam and truss structures are not very well suited for layer by layer fabrication, as the beams are sliced into many small cross-sections. Mueller et al. [2014] propose a dedicated approach to print wire-mesh structures, exploiting the fact that extruded filament hardens quickly to print truss-like structures in mid-air. This raises many challenges regarding path planning of the extrusion device. Wu et al. [2016c] address some of these challenges using a 5DOF printer (see Figure 4.13): two additional degrees of freedom allow the platform to rotate. The path planning problem is formulated as an ordering of the edges of a graph that captures the spatial constraints of extruder motions. Huang et al. [2016] address a similar problem, using a 6DOF robotic arm with a customized extrusion head. In addition to collision avoidance their algorithm considers the stability of the printed part during the whole fabrication process, ensuring that each layer is in a stable equilibrium over its preceding layers.

With the objective of printing hair, fibers, and bristles, Laput et al. [2015] introduced a technique that exploits the stringing phenomena of filament deposition technologies.

Multiple materials.   Some technologies afford for multi-material fabrication. Weiss et al. [1997] describe a process for fabricating multi-material objects. Kumar and Dutta [1998], Kumar et al. [1998] describe a modeling representation and adaptive slicing algorithm for multi-material objects. Zhu and Yu [2001] propose a dexel slicer for multi-material objects. The solid ray intervals are used as solidification paths. Park et al. [2001] describe a system for modeling and fabrication of multi-material objects inspired by procedural volume texturing approaches in computer graphics, noting the advantages for the process planning in terms of memory compactness and resolution independence. Shin et al. [2003] focus on process planning for fabricating objects with continuously varying material properties on a direct metal deposition system.

Machines based on filament extrusion can mount multiple extrusion heads, in which case motion planning during deposition becomes more complex. Choi and Cheung [2005] input a

Figure 4.13: Printed 3D wireframes, using a 5DOF printer. From Wu et al. [2016c]. Image courtesy of Rundong Wu.

triangle mesh per material (packed in a single-colored STL file) that are sliced independently. The contours in each layer are grouped by inclusion order, so as to reduce redundant motions during deposition. This is later extended to motion planning for multiple, independent deposition devices. The challenge is to coordinate the movements of the nozzle depositing different materials, so as to avoid collisions or even deadlocks [Choi and Cheung, 2006, Choi and Zhu, 2010].

In the context of computer graphics, Hergel and Lefebvre [2014] also consider the case of slicing and toolpath planning for multiple materials. Instead of inputing multiple meshes, materials are selected by a *slice shader* which is executed on every point of each layer. Contours are extracted using an image space approach. The motion planner is optimized for visual quality, hiding potential defects in less visible regions of the part. Reiner et al. [2014] achieve a visual grading of colors on multi-filament printers, interleaving the deposited filaments in sine wave patterns along the surface (see Figure 4.14). Kuipers et al. [2017] achieve a similar effect, pushing more or less plastic flow on layers, which make them protrude more or less on side surfaces, while varying the degrees of offsetting used on slanted surfaces. Song and Lefebvre [2017] rely on microlayers and slightly translucent filaments pushed into a mixing nozzle, achieving a wider range of colors and precisely controlled gradients between base filament colors.

Some technologies apply colors on entire layers (inkjet on powder [Cima et al., 1994] and inkjet on laminated paper [Mcor, 2005]). At slicing time, a raster RGB color image is applied to the layer contour. Since the colors are often specified from the surface (e.g., vertex colors or 2D texture map), they are propagated inside from the contour within a thin shell. However, the technology would be able to color the entire volume, even though material opacity limits the potential.

Multi-jet technologies—the deposit droplets of different resins—have a wider variety of materials and colors. The previously mentioned OpenFAB [Vidimče et al., 2013] allows to model complex multi-material geometries with voxels, and streams slices to a high-resolution

Figure 4.14: Producing continuous tone imagery by color mixing on multi-filament printers. From Reiner et al. [2014]. Image courtesy of Rim Reiner.

multi-jet printer. This forms the basis of a complete modeling system for multi-material modeling [Vidimce et al., 2016]. Wu et al. [2000] and Cho et al. [2001] discuss process planning to convert continuously varying material information into a limited set of base materials by a half-toning technique. Brunton et al. [2015] also rely on half-toning in the context of color reproduction—to the point that print-out of scanned objects can be confused with their model. Babaei et al. [2017] layer small patches of colored inks within stacks normal to the object surface. By carefully selecting the thickness of each color layer in each stack they achieve accurate color reproduction, exploiting translucency instead of half-toning.

Earlier work also studied how to control optical properties of the final print by combining multiple base materials [Dong et al., 2010, Hašan et al., 2010]—even though these approaches are not integrated within the process planning pipeline.

4.6   MACHINE INSTRUCTIONS

When talking about machine instructions it is necessary to distinguish between machines that operate on each slice like a plotter, that is, connecting pairs of points with straight lines, and machines that operate on each slice like an inkjet printer, that is, interpreting the whole slice as a discrete 2D image. The former require the slices to be defined in vector format, whereas the latter require the slices to be defined in raster format (see also Section 2.1).

4.6.1   VECTOR CASE

In the vector case the machine instructions amount to a piecewise linear toolpath (in some cases arcs) along which the printer must deposit, melt or sinter the printing material. The machine toolpath must be prepared for each slice, both for its outer contour and the interior. If the part comes in the form of a boundary representation (e.g., a STL file), particular attention must be paid to distinguish between the interior and the exterior of the shape [Volpato et al., 2013]. The generation of a machine toolpath for AM has a clear analogy with CNC pocket milling, where the material inside an arbitrarily closed boundary on a flat surface is removed following a fixed path [Held, 1991, Makhanov and Anotaipaiboon, 2007]. In the general case a machine toolpath intended for milling is, however, not suitable for AM. In FDM the deposition of material along the path poses some additional challenges. The path has to be designed in such a way that the deposition of material is as regular as possible, thus avoiding under and over deposition. For the same reason the overlap between adjacent paths needs a much finer control, in order to cover the slice area with a uniform layer of deposited material [Han et al., 2003] (Figure 4.15). In powder bed technologies such as laser sintering and melting the heat control is fundamental to guarantee quality results. Many printers generate multiple melting pools at a time, in order to better distribute heat and avoid huge thermal gradients, that would generate rough surfaces and warping [Ding et al., 2014]. We discuss here the requirements for a quality machine toolpath and the major differences between available algorithms and technologies.

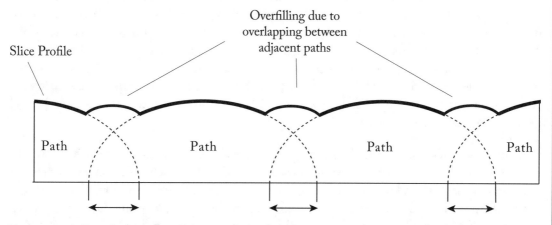

Figure 4.15: Rapid prototyping forming principle [Livesu et al., 2017]. Used with permission.

Continuity. In FDM and other material deposition processes it is important to keep the amount of material deposited along the path constant. To this end, many authors observed that controlling the amount of material being deposited when a path begins or ends is very difficult. Recent works aim to reduce as much as possible the number of disconnected paths which are necessary to cover the slice area, so as to minimize discontinuities in the deposition process [Zhao et al., 2016a].

Geometry. Not only the endpoints of a machine path but also the smoothness of the path itself may affect the quantity of material locally deposited in FDM. Long and low curvature paths are to be preferred to short paths with sharp turns [Jin et al., 2017, Zou et al., 2014]. In the latter case the speed of the nozzle would decrease, thus increasing the time necessary to complete the deposition and possibly triggering under or overfilling of the filament [Jin et al., 2014]. Long and smooth paths are not always possible to generate. If the shape contains small details with sharp creases, short deposition paths with abrupt changes of direction are unavoidable. Žarko et al. [2017] show that small protuberances with regular cross-section (e.g., linear, squared, circular) cannot be faithfully printed with FDM, also discussing how deposition speed affects deviation from the nominal geometry.

Patterns.   Internal volumes of parts are a major factor in time and material consumption, in particular for FDM where the motion of the print head—a relatively heavy mechanical device—is slow. To save time and material, several infill patterns have been proposed. These patterns are usually specific to FDM due to the aforementioned continuity and geometry requirement, but also due to the overhang constraints. See Section 4.4.2 for an in-depth discussion of sparse infill patterns.

Domain split.   In order to be able to process arbitrarily complex slices, most of the approaches proposed in the literature use a divide-and-conquer strategy, partitioning the slice into a set of pockets to operate on, rather than trying to cover the whole area with a single connected curve. To this end, many different strategies have been proposed in the literature. In Dwivedi and Kovacevic [2004], the slice is decomposed into a set of monotone polygons; Ding et al. [2014] use convex decomposition; Zhao et al. [2016a] use the iso-contours of an inward distance field from the border; Held et al. [1994] use a Voronoi-based approach; and Kao and Prinz [1998] use a medial axis based approach.

Performances.   Besides the method used to decompose the domain and the particular curves used to fill each pocket, another important factor in the definition of an efficient toolpath is the reduction of the so-called machine *airtime*, that is, the time necessary to move the nozzle (or the laser) from the end of a curve to the beginning of the subsequent one. Given a toolpath composed of a set of disconnected curves, the machine airtime can be minimized by acting on to three separate variables: the order in which the curves are processed, the path that takes the nozzle from the end of a curve to the beginning of the next one, and the orientation of each curve (from begin to end, or from end to begin). For the latter, notice that for closed loops, the machine may choose any point within the loop as a starting point. The path planning optimization problem has been shown to be related to the Travelling Salesman Problem (TSP), which is NP-complete. Many methods have been proposed for its approximate solution, using genetic algorithms [Wah et al., 2002, Weidong, 2009], the Christofides algorithm [Fok et al., 2016, Ganganath et al., 2016], or other heuristics [Castelino et al., 2003].

4.6.2   RASTER CASE

Recent printing technologies such as Z-Corp. and the DLP/SLA printers (i.e., stereolithography printers that exploit the Digital Light Processing technology) treat each slice as an array of pixels. For the Z-Corp. case the machine toolpath consists in a trivial visit of a regular 2D grid. From a machine usage point of view an important parameter is the binder saturation, which affects both the strength and the accuracy of a printed part [Vaezi and Chua, 2011]. Typically, a binder is first applied with a higher saturation to the edges of the part, creating a strong *shell* for the exterior. Next, an infrastructure is created for the part walls, which are also built with a higher saturation. The remaining interior areas are printed with a lower binder saturation, which gives the part its stability [ZCorporation, 2007]. For the DLP-based printers there is no notion of machine toolpath, because the whole image/slice is projected onto the first layer of the resin tank. In this case it is important to project each slice for the proper amount of time (exposure time), in order to balance surface accuracy with part strength. Exposure time depends on material, layer thickness and slice resolution. Notice that commercial slicers allow to export vector slices (e.g., in SVG format) even for raster printers [Slicer, 2011]. This is because slice rescaling is usually performed internally prior to projection, and scaling a raster image may introduce unnecessary artifacts such as image blurring.

4.7   RELATIONS BETWEEN QUALITY METRICS AND BASIC STEPS

Each of the process planning stages discussed so far has an impact on some of the quality metrics. Since the relative importance of these metrics depends on the application, herewith we provide an overview of these process-quality relations that may help in the process of tuning the whole process planning. Table 4.2 summarizes these relations and, for each of the specific PP steps, refers to the corresponding section where the causes and effects are described in detail.

Note that in many cases several quality metrics should be optimized at the same time, and that is why multiple decision systems and genetic algorithms have been proposed to support the user in the difficult task of finding a tradeoff [Anitha et al., 2001, Ingrassia et al., 2017].

Table 4.2: Relations between process planning steps and quality metrics

	Orientation	Supports	Slicing	Toolpath
Cost				
Pre-build	-	-	-	Section 4.6.1
Build	Section 4.3.1	Section 4.4.2	Section 4.5.1	Section 4.6.1
Post-Processing	Section 4.3.2	Section 4.4.1	-	-
Fidelity				
Form	Section 4.3.2	-	Section 4.5.1	-
Texture	Section 4.3.2	Section 4.4.1	-	-
Design Compliancy	-	-	-	-
Functionality				
Robustness	Section 4.3.3	-	-	-
Mass distribution	-	Section 4.4.2	-	-
Thermal/Mechanical property	-	Section 4.4.2	-	-

CHAPTER 5

Open Challenges

In this final chapter we discuss some open challenges related to design and process planning for AM. We focus on the complementarity of AM with traditional subtractive manufacturing techniques, the open challenges in design for AM and new topics such as the possibility to embed devices into prints and producing self-assembling structures. Finally, we discuss about the needs of industry as far as PP is concerned and the open challenge of a discretization-free pipeline.

5.1 COMPLEMENTARITY WITH SUBTRACTIVE TECHNOLOGIES

In the majority of the mechanically demanding applications, AM is nowadays coupled to subtractive manufacturing. In fact, AM technologies are not capable yet of achieving the precision often required in mechanical parts, especially in terms of features positioning, shape and surface finish, as illustrated in Figure 5.1. In order to address this issue, the common practice is to modify the CAD model leaving some extra material on the features that require high precision (e.g., couplings, holes, pins, shoulders for bearings, etc.). This extra material will be then milled until the desired shape is obtained with the required surface finish. Thanks to this approach, the parts manufactured by means of AM can be used effectively and coupled to traditional parts and commercial components, with a suitable level of precision. Some recent works in process planning integrate both technologies together, e.g. Newman et al. [2015].

Nevertheless, the current limitation in the precision, stability and roughness of the features produced by AM is expected to be solved in the next years of technology development. Similarly to 5-axes milling, which was considerably improved with the development of CNC technologies and with the evolution of CAM tools, it is expected that soon AM will be able to produce parts perfectly meeting the stringent requirements of mechanical couplings.

This would be a double advantage, because not only it will simplify the production of parts and reduce the manufacture cost, but it would allow the users to fully exploit the freedom of AM, overcoming any constraint imposed by the finishing of features by subtractive manufacturing.

5.2 CHALLENGES IN DESIGN

As can be seen in this book, the field of AM gives rise to an abundance of novel or revised applications, most of which have probably not been conceived yet. This leap forward has left a

Figure 5.1: Example of mechanical component (a bevel gear) manufactured with AM (left) and with subtractive technique (right). The surface finish achieved by AM is not compatible with the specifications of the gear, so a finishing operation (e.g., by milling) is required.

gap with respect to design and representation technologies and methodologies. To fully realize the potential of new manufacturing technologies, a fundamental change in design concepts will probably need to happen. Traditional design and representation approaches assume smoothness wherever not explicitly specified otherwise (e.g., the linear interpolation of triangles). This is reasonable when added details induce added cost. This, of course, is not the case for the afore-mentioned technologies, where practically arbitrary details can be produced with no tooling. Therefore, it is likely that hierarchical or programmatic representations, seen in recent publications, will become more popular in the future.

One of the biggest problems of designing in great detail is the limited complexity gras-pable by humans. Therefore, exploring the full design space is not a feasible approach. A possible solution is designing through goals. Ideally, specifying the desired objectives, their trade-offs and constraints, should be sufficient to synthesize the desired design. In this scenario, the designer will have many objectives to choose from, and the difficult task would be to balance them cor-rectly. This will alleviate the need for the designer to grasp all the necessary details of a valid design.

For such a system to be useful, many objectives will probably need to be developed. In this book, some objectives have been introduced, with methods to satisfy them (e.g., articulation by description, deformation behavior, balance, aerodynamics, etc.). These objectives can be dictated by professionals, or deduced from large collections. Some of these objectives are more difficult to manage. Aesthetics, for example, is a major concern for almost any design, but a very elusive one. While this mainly concerns perceptual studies, some quantification for it has been attempted for applications such as integrity control, shape approximation, articulation, and architecture.

These were addressed by heuristics, measuring symmetries, saliency, smoothness, resemblance, etc. A pressing issue would be to further explore these quantifications, enabling better guidance to optimization processes.

Another such objective is motion. This objective can be rather intuitively specified through motion curves or designated constructs such as rigs and kinematic chains. Realizing it, however, is more challenging. Throughout this book, motion was realized through the use of primitives, such as joints, gears, and pulleys, which were manually pre-configured. In the case of deformation control, primitives were also employed, to provide more flexibility. In some approaches, however, these were automatically generated and explored, rather than manually defined beforehand. The notion of exploiting reusable motion and deformation primitives is a promising one, especially as manufacturing resolutions increase. It would be interesting to investigate whether this concept can be applied to other objectives, and whether primitives could be automatically evolved for them. Note that solutions to these problems would also have to address practical aspects, such as tiling the elements in a non-restrictive manner, representing them efficiently, and simulating a design with many instantiations.

As the number of objectives increases, one challenging task designers will face is determining their trade-offs. To date, however, most publications address and solve a specific goal, rather than combining multiple objectives and balancing them. Problems relating more objectives, and their combinations, will probably be tackled as the field matures.

A tightly related question is that of control. Exposing too much control risks overwhelming the designer, but restricting it also risks the design's expressiveness. Typically, novice users require as much automation as possible, while experts would benefit from even direct control at times. Most solutions presented in computer graphics in general, and this book specifically, are inclined for maximal automation, and hence are more fitting for novice users. On the other hand, CAD methods typically aim for professionals, and hence offer more control, but lack high-level optimizations. An open avenue of research is bridging this gap, allowing different levels of control, suited for both novice and professional users.

In addition, some manufacturing technologies were not addressed at all by the computer graphics community. For example, some of the sturdiest designs are those manufactured with metal, and yet metal sintering is not treated with the same level of detail as other techniques. Composite materials also present phenomenal strength-to-weight ratios. In these cases, brittle yet strong reinforcement fibers are embedded in the object, making it significantly more resistant. Fiber-laying strategies likely stand to benefit from geometry processing techniques developed in the realm of computer graphics, and hence also pose an interesting research avenue. Manufacturing by weaving poses great representational challenges, since enormous numbers of features can be concentrated in very small objects. None of these technologies have been addressed by the graphics community so far, but as their popularity grows, they are likely to be the subjects of research in the near future.

5.3   EMBEDDING DEVICES INTO PRINTS

A fascinating and rapidly developing topic is the ability to insert sensors or devices into a print. For instance, Savage and colleagues proposed a design tool to augment objects with specially fabricated touch sensitive areas [Savage et al., 2012], and to embed a camera into 3D printed objects to turn them into custom controllers [Savage et al., 2013]. The RevoMaker [Gao et al., 2015] fabricates an object *around* a core containing sensors and electronic devices. The SurfCuit [Umetani and Schmidt, 2017] allows the user to interactively integrate circuits into 3D prints. The *Voxel8* printer goes beyond this by proposing to integrate electronic components connected together by conductive ink deposited within the layers. Similarly, Peng et al. [2016] describe a prototype printer that, besides depositing fused plastic, can also roll copper coils that produce magnetic fields when traversed by a current. Such conductive materials open a wide range of applications, in particular for antenna or battery designs [Raney and Lewis, 2015]. The embedding of antennas aimed at the generation of 3D printed internet connected devices is the subject of Iyer et al. [2017]. The authors demonstrate a wide range of applications including sensing, gaming, and tracking. These applications open a whole new set of challenges overlapping between material science, geometry, 3D modeling, circuit routing, and process planning for AM.

5.4   SELF-ASSEMBLING STRUCTURES

Another promising direction of research is the study of materials that are able to morph into a target shape after being printed, the so-called 4D printing [Tibbits, 2014]. For instance, to print plant-inspired structures that change their shape after immersion in water [Gladman et al., 2016]. Self-assembly structures have applications in autonomous robotics or tissue engineering, among others [Khoo et al., 2015].

5.5   CHALLENGES IN AN INDUSTRIAL PERSPECTIVE

AM has a huge potential for industry, thanks to the many advantages introduced by the technology itself and also the extreme simplification of the process planning compared to traditional approaches. Nevertheless, several challenges still have to be solved to enable a large adoption of AM by the industry.

Industry typically seeks integrated solutions for PP, i.e., algorithms for model repair, part orientation, support creation, slicing, toolpath calculation etc. shall be preferably integrated in a single, comprehensive tool. In this regard, one of the key challenges is to keep such tools up to date and allow a suitable degree of customization. At the academic level, the development of a common software framework for AM would help toward this goal: by developing their research prototypes within a common framework, academics could reach a larger audience and ease transfer toward the industry.

An interesting trend is to attempt to automatically optimize most steps of the process (part orientation, supports, batch printing, etc.). However, it is important to give enough controls to users. For instance some users will prioritize the performance of the part (structural resistance, surface finish, etc.), while other users would be interested in optimizing the process (production time, material waste, etc.). To this end, the parameter(s) to be optimized by PP algorithms must be selected by the user. Unlike for non-professional users, tools where the parameters cannot be tuned would not be acceptable.

Another area of improvement of PP is the process simulation. As we have seen in this book, some key characteristics of the final result can already be reliably forecast and some of the features can now be subjected to an optimization loop. The future development will hopefully produce more and more accurate tools that can be integrated in the PP, allowing the users to improve desiderata and minimize the uncertainties related to the AM technology.

5.6 DISCRETIZATION-FREE PIPELINE

As discussed earlier in this book, the usual entry point for Process Planning consists of a tessellation that approximates the smoothly curved CAD model with a collection of triangles. This step simplifies geometric operations and exchanges between software to the point that the triangle-based STL format has become a de-facto standard for process planning. Nonetheless, since the tessellation step is often the cause of many issues with the geometry (see Section 4.1.2) and introduces unnecessary approximations, many attempts have been made during the years to avoid this conversion and perform all the calculations based on the original CAD geometry (see Section 4.5.2). In spite of these efforts, STL remains an undiscussed standard and CAD-compliant geometric processing for PP is still a very limited practice.

A main obstacle to a widespread adoption of such pipelines is the intrinsic hardware limitations of current printing devices. 3D printers, indeed, can just execute a finite set of commands that typically makes the tool move from one point to another along a straight line segment or, in the most sophisticated models, to follow circular arcs while extruding or solidifying material.

Nevertheless, a discretization-free pipeline would allow a form of device independent process planning, where discretization would only occur at the very last moment, on the machine. Thus, the approximations would be optimally decided depending upon machine capabilities and resolution.

Bibliography

Adhikary and Gurumoorthy. Direct global editing of STL mesh model for product design and rapid prototyping. *Rapid Prototyping Journal*, 23(4):781–795, 2017. DOI: 10.1108/rpj-06-2015-0064. 55

Daekeon Ahn, Hochan Kim, and Seokhee Lee. Fabrication direction optimization to minimize post-machining in layered manufacturing. *International Journal Machine Tools Manufacture*, 47(3):593–606, 2007. DOI: 10.1016/j.ijmachtools.2006.05.004. 61

Marc Alexa, Kristian Hildebrand, and Sylvain Lefebvre. Optimal discrete slicing. *ACM Transactions on Graphics (TOG)*, 36(1):12, 2017. DOI: 10.1145/3072959.3126803. 76

Paul Alexander, Seth Allen, and Debasish Dutta. Part orientation and build cost determination in layered manufacturing. *Computer-Aided Design*, 30(5):343–356, 1998. DOI: 10.1016/s0010-4485(97)00083-3. 10, 12, 60

Andrew Allen and Nikunj Raghuvanshi. Aerophones in flatland: Interactive wave simulation of wind instruments. *ACM Transactions on Graphics*, 34(4):134:1–134:11, July 2015. http://doi.acm.org/10.1145/2767001 DOI: 10.1145/2767001. 35

Robert J. A. Allen and Richard S. Trask. An experimental demonstration of effective curved layer fused filament fabrication utilising a parallel deposition robot. *Additive Manufacturing*, 8:78–87, 2015. DOI: 10.1016/j.addma.2015.09.001. 80

Seth Allen and Deba Dutta. On the computation of part orientation using support structures in layered manufacturing. In *Proc. of Solid Freeform Fabrication Symposium*, pages 259–269, 1994. 60

Seth Allen and Deba Dutta. Determination and evaluation of support structures in layered manufacturing. *Journal of Design and Manufacturing*, 5:153–162, 1995. 66

Joseph W. Allison, Thomas P. Chen, Adam L. Cohen, Dennis R. Smalley, David E. Snead, and Thomas J. Vorgitch. Boolean layer comparison slice. *U.S. Patent 5854748*, 3D Systems Inc., 1988. 66

Erik Andreassen, Boyan S. Lazarov, and Ole Sigmund. Design of manufacturable 3D extremal elastic microstructure. *Mechanics of Materials*, 69(1):1–10, 2014. DOI: 10.1016/j.mechmat.2013.09.018. 74

Anitha, Arunachalam, and Radhakrishnan. Critical parameters influencing the quality of prototypes in fused deposition modelling. *Journal of Materials Processing Technology*, 118(1):385–
388, 2001. DOI: 10.1016/s0924-0136(01)00980-3. 85

Marco Attene. A lightweight approach to repair polygon meshes. *The Visual Computer*,
pages 1393–1406, 2010. DOI: 10.1007/s00371-010-0416-3. 56, 57

Marco Attene. Direct repair of self-intersecting meshes. *Graphical Models*, 76:658–668, 2014.
DOI: 10.1016/j.gmod.2014.09.002. 56

Marco Attene. Shapes in a box: Disassembling 3D objects for efficient packing and fabrication.
Computer Graphics Forum, 34(8):64–76, 2015. DOI: 10.1111/cgf.12608. 60

Marco Attene. As-exact-as-possible repair of unprintable STL files. *Rapid Prototyping Journal*,
24(7), (to appear). 56

Marco Attene, Marcel Campen, and Leif Kobbelt. Polygon mesh repairing: An
application perspective. *ACM Computing Surveys*, 45(2):15:1–15:33, 2013. DOI:
10.1145/2431211.2431214. 56

Autodesk. Meshmixer. http://www.meshmixer.com, 2011. 56

Vahid Babaei, Kiril Vidimče, Michael Foshey, Alexandre Kaspar, Piotr Didyk, and Wojciech
Matusik. Color contoning for 3D printing. *ACM Transactions on Graphics*, 36(4):124:1–
124:15, July 2017. DOI: 10.1145/3072959.3073605. 82

Moritz Bächer, Bernd Bickel, Doug L. James, and Hanspeter Pfister. Fabricating articulated
characters from skinned meshes. *ACM Transactions on Graphics*, 31(4):47:1–47:9, July 2012.
DOI: 10.1145/2185520.2185543. 30, 48, 52

Moritz Bächer, Emily Whiting, Bernd Bickel, and Olga Sorkine-Hornung. Spin-It: Optimizing moment of inertia for spinnable objects. *ACM Transactions on Graphics*, 33(4):96:1–96:10,
2014. DOI: 10.1145/2601097.2601157. 14, 34, 47, 48

Moritz Bächer, Stelian Coros, and Bernhard Thomaszewski. Linkedit: Interactive linkage editing using symbolic kinematics. *ACM Transactions on Graphics*, 34(4):99:1–99:8, July 2015.
DOI: 10.1145/2766985. 31

Ilya Baran, Philipp Keller, Derek Bradley, Stelian Coros, Wojciech Jarosz, Derek
Nowrouzezahrai, and Markus Gross. Manufacturing layered attenuators for multiple prescribed shadow images. *Computer Graphics Forum (Proceedings of Eurographics)*, 31(2):603–
610, May 2012. DOI: 10.1111/j.1467-8659.2012.03039.x. 17

Mohand Ourabah Benouamer and Dominique Michelucci. Bridging the gap between CSG
and Brep via a triple ray representation. In *Proc. of the 4th ACM Symposium on Solid Modeling
and Applications*, 1997. DOI: 10.1145/267734.267755. 78

Amit Bermano, Ilya Baran, Marc Alexa, and Wojciech Matusk. Shadowpix: Multiple images from self shadowing. *Computer Graphics Forum*, 31(2pt3):593–602, May 2012. DOI: 10.1111/j.1467-8659.2012.03038.x. 17

Amit Haim Bermano, Thomas Funkhouser, and Szymon Rusinkiewicz. State of the art in methods and representations for fabrication-aware design. *Computer Graphics Forum (Eurographics STAR)*, 36(2), 2017. DOI: 10.1111/cgf.13146. 2

James M. Bern, Kai-Hung Chang, and Stelian Coros. Interactive design of animated plushies. *ACM Transactions on Graphics*, 36(4):80:1–80:11, July 2017. DOI: 10.1145/3072959.3073700. 40

Gaurav Bharaj, Stelian Coros, Bernhard Thomaszewski, James Tompkin, Bernd Bickel, and Hanspeter Pfister. Computational design of walking automata. In *Proc. of the 14th ACM SIGGRAPH/Eurographics Symposium on Computer Animation, (SCA'15)*, pages 93–100, 2015. DOI: 10.1145/2786784.2786803. 31, 52

Bernd Bickel, Moritz Bächer, Miguel A. Otaduy, Hyunho Richard Lee, Hanspeter Pfister, Markus Gross, and Wojciech Matusik. Design and fabrication of materials with desired deformation behavior. *ACM Transactions on Graphics*, 29(4):1, 2010. DOI: 10.1145/1833351.1778800. 24, 26, 47, 52

Bernd Bickel, Peter Kaufmann, Mélina Skouras, Bernhard Thomaszewski, Derek Bradley, Thabo Beeler, Phil Jackson, Steve Marschner, Wojciech Matusik, and Markus Gross. Physical face cloning. *ACM Transactions on Graphics*, 31(4):118:1–118:10, July 2012. DOI: 10.1145/2185520.2335469. 25, 47, 50, 52

Stephan Bischoff, Darko Pavic, and Leif Kobbelt. Automatic restoration of polygon models. *ACM Transactions on Graphics*, 24(4):1332–1352, 2005. DOI: 10.1145/1095878.1095883. 56

D. Brackett, I. Ashcroft, and R. Hague. Topology optimization for additive manufacturing. In *Proc. of the Solid Freeform Fabrication Symposium*, pages 348–362, 2011. 15

Alan Brunton, Can Ates Arikan, and Philipp Urban. Pushing the limits of 3D color printing: Error diffusion with translucent materials. *ACM Transactions on Graphics*, 37(1):1–13, 2015. DOI: 10.1145/2832905. 18, 47, 48, 52, 82

Hong-Seok Byun and Kwan H. Lee. Determination of optimal build direction in rapid prototyping with variable slicing. *International Journal of Advanced Manufacturing Technology*, 28 (3-4):307–313, 2006a. DOI: 10.1007/s00170-004-2355-5. 10, 63

Hong-Seok Byun and Kwan H. Lee. Determination of the optimal build direction for different rapid prototyping processes using multi-criterion decision making. *Robotics and Computer-Integrated Manufacturing*, 22(1):69–80, 2006b. DOI: 10.1016/j.rcim.2005.03.001. 10, 63

Daniela Cabiddu and Marco Attene. Epsilon-maps: Characterizing, detecting and thickening thin features in geometric models. *Computers and Graphics*, 66(Supplement C):143–153, 2017. Shape Modeling International 2017. DOI: 10.1016/j.cag.2017.05.014. 55

Simone Cacace, Emiliano Cristiani, and Leonardo Rocchi. A level set based method for fixing overhangs in 3D printing. *Applied Mathematical Modelling*, 44:446–455, 2017. DOI: 10.1016/j.apm.2017.02.004. 63

Joseph E. Cadman, Shiwei Zhou, Yuhang Chen, and Qing Li. On design of multifunctional microstructural materials. *Journal of Materials Science*, 48(1):51–66, 2013. DOI: 10.1007/s10853-012-6643-4. 73

Stéphane Calderon and Tamy Boubekeur. Point morphology. *ACM Transactions on Graphics*, 33(4):45:1–45:13, 2014. DOI: 10.1145/2601097.2601130. 69

Jacques Calì, Dan A. Calian, Cristina Amati, Rebecca Kleinberger, Anthony Steed, Jan Kautz, and Tim Weyrich. 3D-printing of non-assembly, articulated models. *ACM Transactions on Graphics*, 31(6):130:1–130:8, November 2012. DOI: 10.1145/2366145.2366149. 30, 52

Flaviana Calignano. Design optimization of supports for overhanging structures in aluminum and titanium alloys by selective laser melting. *Materials and Design*, 64:203–213, 2014. DOI: 10.1016/j.matdes.2014.07.043. 67

Marcel Campen and Leif Kobbelt. Exact and robust (self-)intersections for polygonal meshes. *Computer Graphics Forum*, 29(2):397–406, 2010a. DOI: 10.1111/j.1467-8659.2009.01609.x. 56

Marcel Campen and Leif Kobbelt. Polygonal boundary evaluation of Minkowski sums and swept volumes. *Computer Graphics Forum*, 29(5):1613–1622, 2010b. DOI: 10.1111/j.1467-8659.2010.01770.x. 68

Vassilios Canellidis, Vassilis Dedoussis, N. Mantzouratos, and S. Sofianopoulou. Pre-processing methodology for optimizing stereolithography apparatus build performance. *Computers in Industry*, 57(5):424–436, 2006. DOI: 10.1016/j.compind.2006.02.004. 60

Loren Carpenter. The A-buffer, an antialiased hidden surface method. In *Proc. of SIGGRAPH*, pages 103–108, 1984. DOI: 10.1145/964965.808585. 77

Thomas J. Cashman. Beyond catmull-clark? A survey of advances in subdivision surface methods. *Computer Graphics Forum*, 31(1):42–61, 2012. DOI: 10.1111/j.1467-8659.2011.02083.x. 50

Kenneth Castelino, Roshan D'Souza, and Paul K. Wright. Toolpath optimization for minimizing airtime during machining. *Journal of Manufacturing Systems*, 22(3):173–180, 2003. DOI: 10.1016/s0278-6125(03)90018-5. 84

Duygu Ceylan, Wilmot Li, Niloy J. Mitra, Maneesh Agrawala, and Mark Pauly. Designing and fabricating mechanical automata from mocap sequences. *ACM Transactions on Graphics*, 32(6):186:1–186:11, November 2013. DOI: 10.1145/2508363.2508400. 31, 52

Debapriya Chakraborty, B. Aneesh Reddy, and A. Roy Choudhury. Extruder path generation for curved layer fused deposition modeling. *Computer-Aided Design*, 40(2):235–243, 2008. DOI: 10.1016/j.cad.2007.10.014. 80

Kumar Chalasani, Larry Jones, and Larry Roscoe. Support generation for fused deposition modeling. In *Solid Freeform Fabrication Symposium*, pages 229–241, 1995. 66

Ian Chan, William Wells III, Robert V. Mulkern, Steven Haker, Jianqing Zhang, Kelly H. Zou, Stephan E. Maier, and Clare M. C. Tempany. Detection of prostate cancer by integration of line-scan diffusion, t2-mapping and t2-weighted magnetic resonance imaging; a multichannel statistical classifier. *Medical Physics*, 30(9):2390–2398, 2003. DOI: 10.1118/1.1593633. 47

Angel X. Chang, Thomas Funkhouser, Leonidas Guibas, Pat Hanrahan, Qixing Huang, Zimo Li, Silvio Savarese, Manolis Savva, Shuran Song, Hao Su, Jianxiong Xiao, Li Yi, and Fisher Yu. ShapeNet: An information-rich 3D model repository. *Technical Report arXiv:1512.03012* [cs.GR], Stanford University, Princeton University, Toyota Technological Institute at Chicago, 2015. 38

Desai Chen, David I. W. Levin, Piotr Didyk, Pitchaya Sitthi-Amorn, and Wojciech Matusik. Spec2fab: A reducer-tuner model for translating specifications to 3D prints. *ACM Transactions on Graphics*, 32(4):1, 2013a. DOI: 10.1145/2461912.2461994. 8, 42, 52

Desai Chen, David I. Levin, Wojciech Matusik, and Danny M. Kaufman. Dynamics-aware numerical coarsening for fabrication design. *ACM Transactions on Graphics*, 34(4), 2017a. DOI: 10.1145/3072959.3073669. 25

Lujie Chen and Lawrence Sass. Fresh press modeler: A generative system for physically based low fidelity prototyping. *Computers and Graphics*, 54:157–165, 2016. Special Issue on CAD/Graphics, 2015. DOI: 10.1016/j.cag.2015.07.003. 42, 49

Weikai Chen, Xiaolong Zhang, Shiqing Xin, Yang Xia, Sylvain Lefebvre, and Wenping Wang. Synthesis of filigrees for digital fabrication. *ACM Transactions on Graphics*, 35(4):98:1–98:13, July 2016. DOI: 10.1145/2897824.2925911. 20, 21, 48, 49, 52

Weikai Chen, Yuexin Ma, Sylvain Lefebvre, Shiqing Xin, Jonàs Martínez, and Wenping Wang. Fabricable tile decors. *ACM Transactions on Graphics (TOG)*, 36(6):XX, 2017b. DOI: 10.1145/3130800.3130817. 20

Xiang Chen, Changxi Zheng, Weiwei Xu, and Kun Zhou. An asymptotic numerical method for inverse elastic shape design. *ACM Transactions on Graphics*, 33(4):95:1–95:11, July 2014. DOI: 10.1145/2601097.2601189. 27, 47, 52

Xuelin Chen, Hao Zhang, Jinjie Lin, Ruizhen Hu, Lin Lu, Qixing Huang, Bedrich Benes, Daniel Cohen-Or, and Baoquan Chen. Dapper: Decompose-and-pack for 3D printing. *ACM Transactions on Graphics*, 34(6):213:1–213:12, October 2015. DOI: 10.1145/2816795.2818087. 60

Yong Chen. 3D texture mapping for rapid manufacturing. *Computer Aided Design Applications*, 4(6):761–771, 2007. DOI: 10.1080/16864360.2007.10738509. 74

Yong Chen and Charlie C. L. Wang. Regulating complex geometries using layered depth—normal images for rapid prototyping and manufacturing. *Rapid Prototyping Journal*, 19(4): 253–268, 2013. DOI: 10.1108/13552541311323263. 56, 78

Yong Chen and Charlie C. L. Wang. Layer depth-normal images for complex geometries: Part one—accurate modeling and adaptive sampling. In *International Design Engineering Technical Conferences and Computers and Information in Engineering Conference*, 2008. DOI: 10.1115/detc2008-49432. 69, 77

Yong Chen and Charlie C. L. Wang. Uniform offsetting of polygonal model based on layered depth-normal images. *Computer-Aided Design*, 43(1):31–46, 2011. DOI: 10.1016/j.cad.2010.09.002. 69

Yong Chen, Kang Li, and Xiaoping Qian. Direct geometry processing for telefabrication. *Journal of Computer Information Science in Engineering*, 13(4):041002, 2013b. DOI: 10.1115/1.4024912. 66, 79

W. Cheng, J. Y. H. Fuh, A. Y. C. Nee, Y. S. Wong, H. T. Loh, and T. Miyazawa. Multi-objective optimization of part-building orientation in stereolithography. *Rapid Prototyping Journal*, 1(4):12–23, 1995. DOI: 10.1108/13552549510104429. 60

W. K. Chiu and S. T. Tan. Using dexels to make hollow models for rapid prototyping. *Computer-Aided Design*, 30(7):539–547, 1998. DOI: 10.1016/s0010-4485(98)00008-6. 69

W. Cho, E. M. Sachs, N. M. Patrikalakis, M. J. Cima, T. R. Jackson, H. Liu, J. Serdy, C. C. Stratton, H. Wu, and R. Resnick. Methods for distributed design and fabrication of parts with local composition control. In *Proc. of the NSF Design and Manufacturing Grantees Conference*, 2001. 82

S. H. Choi and H. H. Cheung. A multi-material virtual prototyping system. *Computer-Aided Design*, 37(1):123–136, 2005. DOI: 10.1016/j.cad.2004.06.002. 80

S. H. Choi and H. H. Cheung. A topological hierarchy-based approach to toolpath planning for multi-material layered manufacturing. *Computer-Aided Design*, 38(2):143–156, 2006. DOI: 10.1016/j.cad.2005.08.005. 81

S. H. Choi and W. K. Zhu. A dynamic priority-based approach to concurrent toolpath planning for multi-material layered manufacturing. *Computer-Aided Design*, 42(12):1095–1107, 2010. DOI: 10.1016/j.cad.2010.07.004. 81

Asger Nyman Christiansen, J. Andreas Bærentzen, Morten Nobel-Jørgensen, Niels Aage, and Ole Sigmund. Combined shape and topology optimization of 3D structures. *Computers and Graphics*, 46:25–35, 2015. Shape Modeling International, 2014. DOI: 10.1016/j.cag.2014.09.021. 23, 47, 52

Paolo Cignoni, Nico Pietroni, Luigi Malomo, and Roberto Scopigno. Field-aligned mesh joinery. *ACM Transactions on Graphics*, 33(1):11:1–11:12, February 2014. DOI: 10.1145/2537852. 42, 53

Michael Cima, Emanuel Sachs, Tailin Fan, James F. Bredt, Steven P. Michaels, Satbir Khanuja, Alan Lauder, Sang-Joon J. Lee, David Brancazio, Alain Curodeau, et al. Three-dimensional printing techniques, August 1994. *U.S. Patent 5340656*. 4, 81

Stelian Coros, Bernhard Thomaszewski, Gioacchino Noris, Shinjiro Sueda, Moira Forberg, Robert W. Sumner, Wojciech Matusik, and Bernd Bickel. Computational design of mechanical characters. *ACM Transactions on Graphics*, 32(4):83:1–83:12, July 2013. DOI: 10.1145/2461912.2461953. 31, 52

J. Austin Cottrell, Thomas J. R. Hughes, and Yuri Bazilevs. *Isogeometric Analysis: Toward Integration of CAD and FEA*. John Wiley & Sons, 2009. DOI: 10.1002/bate.201190060. 43

Jordan J. Cox, Yasuko Takezaki, Helaman R. P. Ferguson, Kent E. Kohkonen, and Eric L. Mulkay. Space-filling curves in tool-path applications. *Computer-Aided Design*, 26(3):215–224, 1994. DOI: 10.1016/0010-4485(94)90044-2. 71

S. Scott Crump. Apparatus and method for creating three-dimensional objects. *U.S. Patent 5121329*, 1989. 3

Fernando de Goes, Mathieu Desbrun, Mark Meyer, and Tony DeRose. Subdivision exterior calculus for geometry processing. *ACM Transactions on Graphics*, 35(4):133:1–133:11, July 2016. DOI: 10.1145/2897824.2925880. 51

Ercan M. Dede, Shailesh N. Joshi, and Feng Zhou. Topology optimization, additive layer manufacturing, and experimental testing of an air-cooled heat sink. *Journal of Mechanical Design*, 137(11):111403, 2015. DOI: 10.1115/1.4030989. 15

P. Delfs, M. Tows, and H.-J. Schmid. Optimized build orientation of additive manufactured parts for improved surface quality and build time. *Additive Manufacturing*, 2016. DOI: 10.1016/j.addma.2016.06.003. 13, 61

Mario Deuss, Daniele Panozzo, Emily Whiting, Yang Liu, Philippe Block, Olga Sorkine-Hornung, and Mark Pauly. Assembling self-supporting structures. *ACM Transactions on Graphics*, 33(6):214:1–214:10, November 2014. DOI: 10.1145/2661229.2661266. 49

Donghong Ding, Zengxi Stephen Pan, Dominic Cuiuri, and Huijun Li. A tool-path generation strategy for wire and ARC additive manufacturing. *International Journal of Advanced Manufacturing Technology*, 73(1-4):173–183, 2014. DOI: 10.1007/s00170-014-5808-5. 71, 83, 84

H. Quynh Dinh, Filipp Gelman, Sylvain Lefebvre, and Frédéric Claux. Modeling and toolpath generation for consumer-level 3D printing. In *ACM SIGGRAPH Courses*, pages 17:1–17:273, 2015. DOI: 10.1145/2776880.2792702. 72

André Dolenc and Ismo Mäkelä. Slicing procedures for layered manufacturing techniques. *Computer-Aided Design*, 26(2):119–126, 1994. DOI: 10.1016/0010-4485(94)90032-9. 12, 76

Yue Dong, Jiaping Wang, Fabio Pellacini, Xin Tong, and Baining Guo. Fabricating spatially-varying subsurface scattering. *ACM Transactions on Graphics*, 29(4):1, 2010. DOI: 10.1145/1833351.1778799. 82

Tao Du, Adriana Schulz, Bo Zhu, Bernd Bickel, and Wojciech Matusik. Computational multicopter design. *ACM Transactions on Graphics*, 35(6):227:1–227:10, November 2016. DOI: 10.1145/2980179.2982427. 36

Jérémie Dumas, Jean Hergel, and Sylvain Lefebvre. Bridging the gap: Automated steady scaffoldings for 3D printing. *ACM Transactions on Graphics*, 33(4):98:1–98:10, 2014. DOI: 10.1145/2601097.2601153. 48, 64, 66, 67

Jérémie Dumas, An Lu, Sylvain Lefebvre, Jun Wu, T. U. München, Christian Dick, and T. U. München. By-example synthesis of structurally sound patterns. *ACM Transactions on Graphics*, 34(4):137:1–137:12, July 2015. DOI: 10.1145/2766984. 20, 47, 48, 52, 53

Rajeev Dwivedi and Radovan Kovacevic. Automated torch path planning using polygon subdivision for solid freeform fabrication based on welding. *Journal of Manufacturing Systems*, 23(4):278–291, 2004. DOI: 10.1016/s0278-6125(04)80040-2. 84

Gerald Eggers and Kurt Renap. Method and apparatus for automatic support generation for an object made by means of a rapid prototype production method, 2007. *U.S. Patent 20100228369*, Materialize. 66

Michael Eigensatz, Martin Kilian, Alexander Schiftner, Niloy J. Mitra, Helmut Pottmann, and Mark Pauly. Paneling architectural freeform surfaces. *ACM Transactions on Graphics*, 29(4), Article 45, July 2010. DOI: 10.1145/1778765.1778782. 44, 50

Tawfik T. El-Midany, Ahmed Elkeran, and Hamdy Tawfik. Toolpath pattern comparison: Contour-parallel with direction-parallel. In *Geometric Modeling and Imaging—New Trends*, pages 77–82, 2006. DOI: 10.1109/gmai.2006.45. 71

Oskar Elek, Denis Sumin, Ran Zhang, Tim Weyrich, Karol Myszkowski, Bernd Bickel, Alexander Wilkie, and Jaroslav Křivánek. Scattering-aware texture reproduction for 3D printing. *ACM Transactions on Graphics*, 36(6):241:1–241:15, November 2017. http://doi.acm.org/10.1145/3130800.3130890 DOI: 10.1145/3130800.3130890. 18

Ben Ezair, Fady Massarwi, and Gershon Elber. Orientation analysis of 3D objects toward minimal support volume in 3D-printing. *Computers and Graphics*, 51:117–124, 2015. DOI: 10.1016/j.cag.2015.05.009. 61

Rida T. Farouki. Exact offset procedures for simple solids. *Computer Aided Geometric Design*, 2 (4):257–279, 1985. DOI: 10.1016/s0167-8396(85)80002-9. 68

Kai-Yin Fok, Nuwan Ganganath, Chi-Tsun Cheng, and K. Tse Chi. A 3D printing path optimizer based on Christofides algorithm. In *IEEE International Conference on Consumer Electronics-Taiwan*, pages 1–2, 2016. DOI: 10.1109/icce-tw.2016.7520990. 84

Mark Forsyth. Shelling and offsetting bodies. In *Proc. of the 3rd ACM Symposium on Solid Modeling and Applications*, pages 373–381, 1995. DOI: 10.1145/218013.218088. 68

Dietmar Frank and Georges Fadel. Expert system-based selection of the preferred direction of build for rapid prototyping processes. *Journal of Intelligent Manufacturing*, 6(5):339–345, 1995. DOI: 10.1007/bf00124677. 60

Sarah F. Frisken, Ronald N. Perry, Alyn P. Rockwood, and Thouis R. Jones. Adaptively sampled distance fields: A general representation of shape for computer graphics. In *Proc. of SIGGRAPH*, pages 249–254, 2000. DOI: 10.1145/344779.344899. 68

Oleg Fryazinov, Turlif Vilbrandt, and Alexander A. Pasko. Multi-scale space-variant FRep cellular structures. *Computer-Aided Design*, 45(1):26–34, 2013. DOI: 10.1016/j.cad.2011.09.007. 74

Chi-Wing Fu, Chi-Fu Lai, Ying He, and Daniel Cohen-Or. K-set tilable surfaces. *ACM Transactions on Graphics, (SIGGRAPH)*, 29(3):44:1–44:6, August 2010. DOI: 10.1145/1833349.1778781. 44, 49

Chi-Wing Fu, Peng Song, Xiaoqi Yan, Lee Wei Yang, Pradeep Kumar Jayaraman, and Daniel Cohen-Or. Computational interlocking furniture assembly. *ACM Transactions on Graphics*, 34(4):91:1–91:11, July 2015. DOI: 10.1145/2766892. 38, 52

Nuwan Ganganath, Chi-Tsun Cheng, Kai-Yin Fok, and K. Tse Chi. Trajectory planning for 3D printing: A revisit to traveling salesman problem. In *2nd International Conference on Control, Automation and Robotics*, pages 287–290, 2016. DOI: 10.1109/iccar.2016.7486742. 84

Wei Gao, Yunbo Zhang, Diogo C. Nazzetta, Karthik Ramani, and Raymond J. Cipra. Revomaker: Enabling multi-directional and functionally-embedded 3D printing using a rotational cuboidal platform. In *Proc. of UIST*, pages 437–446, 2015. DOI: 10.1145/2807442.2807476. 90

Julien Gardan. Additive manufacturing technologies: state of the art and trends. *International Journal of Production Research*, 54(10):3118–3132, 2016. DOI: 10.1080/00207543.2015.1115909. 2

Akash Garg, Andrew O. Sageman-Furnas, Bailin Deng, Yonghao Yue, Eitan Grinspun, Mark Pauly, and Max Wardetzky. Wire mesh design. *ACM Transactions on Graphics*, 33(4):66:1–66:12, July 2014. 40, 41, 42, 50, 53

Akash Garg, Alec Jacobson, and Eitan Grinspun. Computational design of reconfigurables. *ACM Transactions on Graphics (TOG)*, 35(4), 2016. DOI: 10.1145/2897824.2925900. 36

Damien Gauge, Stelian Coros, Sandro Mani, and Bernhard Thomaszewski. Interactive design of modular tensegrity characters. In *Proc. of the ACM SIGGRAPH/Eurographics Symposium on Computer Animation, (SCA'14)*, pages 131–138. Eurographics Association, 2014. `http://dl.acm.org/citation.cfm?id=2849517.2849539` DOI: 10.2312/sca.20141131. 29, 52

A. Sydney Gladman, Elisabetta A. Matsumoto, Ralph G. Nuzzo, L. Mahadevan, and Jennifer A. Lewis. Biomimetic 4D printing. *Nature Materials*, 15:413–418, 2016. DOI: 10.1038/nmat4544. 90

John G. Griffiths. Toolpath based on Hilbert's curve. *Computer-Aided Design*, 26(11):839–844, 1994. DOI: 10.1016/0010-4485(94)90098-1. 71

Ruslan Guseinov, Eder Miguel, and Bernd Bickel. Curveups: Shaping objects from flat plates with tension-actuated curvature. *ACM Transactions on Graphics*, 36(4):64:1–64:12, July 2017. DOI: 10.1145/3072959.3073709. 28

Peter Hachenberger, Lutz Kettner, and Kurt Mehlhorn. Boolean operations on 3D selective Nef complexes: Data structure, algorithms, optimized implementation and experiments. *Computational Geometry*, 38(1–2):64–99, 2007. DOI: 10.1016/j.comgeo.2006.11.009. 56

Wenbiao Han, Mohsen A. Jafari, and Kian Seyed. Process speeding up via deposition planning in fused deposition-based layered manufacturing processes. *Rapid Prototyping Journal*, 9(4):212–218, 2003. DOI: 10.1108/13552540310489596. 83

Jingbin Hao, Liang Fang, and Robert E. Williams. An efficient curvature-based partition-ing of large—scale STL models. *Rapid Prototyping Journal*, 17(2):116–127, 2011. DOI: 10.1108/13552541111113862. 58

Eugene E. Hartquist, J. P. Menon, Krishnan Suresh, Herbert B. Voelcker, and Jovan Zagajac. A computing strategy for applications involving offsets, sweeps, and Minkowski operations. *Computer-Aided Design*, 31(3):175–183, 1999. DOI: 10.1016/s0010-4485(99)00014-7. 69

Miloš Hašan, Martin Fuchs, Wojciech Matusik, Hanspeter Pfister, and Szymon Rusinkiewicz. Physical reproduction of materials with specified subsurface scattering. *ACM Transactions on Graphics*, 29(3), 2010. DOI: 10.1145/1778765.1778798. 82

Mohammad T. Hayasi and Bahram Asiabanpour. A new adaptive slicing approach for the fully dense freeform fabrication (FDFF) process. *Journal of Intelligent Manufacturing*, 24(4):683–694, 2013. DOI: 10.1007/s10845-011-0615-4. 76

Erik K. Heide. Method for generating and building support structures with deposition-based digital manufacturing systems, 2010. *U.S. Patent 20110178621 A1*, Stratasys. 66

Martin Held. A geometry-based investigation of the tool path generation for zigzag pocket machining. *The Visual Computer*, 7(5-6):296–308, 1991. DOI: 10.1007/bf01905694. 83

Martin Held, Gábor Lukács, and László Andor. Pocket machining based on contour-parallel tool paths generated by means of proximity maps. *Computer-Aided Design*, 26(3):189–203, 1994. DOI: 10.1016/0010-4485(94)90042-6. 71, 84

Jean Hergel and Sylvain Lefebvre. Clean color: Improving multi-filament 3D prints. *Computer Graphics Forum*, 33(2):469–478, 2014. DOI: 10.1111/cgf.12318. 17, 47, 48, 81

Philipp Herholz, Wojciech Matusik, and Marc Alexa. Approximating free-form geometry with height fields for manufacturing. *Computer Graphics Forum*, 34(2):239–251, 2015. DOI: 10.1111/cgf.12556. 59

Kristian Hildebrand, Bernd Bickel, and Marc Alexa. CRDBRD: Shape fabrication by slid-ing planar slices. *Computer Graphics Forum*, 31(2pt3):583–592, 2012. DOI: 10.1111/j.1467-8659.2012.03037.x. 41, 48, 53

Kristian Hildebrand, Bernd Bickel, and Marc Alexa. Orthogonal slicing for additive manufac-turing. *Computers and Graphics*, 37(6):669–675, 2013. DOI: 10.1016/j.cag.2013.05.011. 47, 48, 58, 61, 76

Owen J. Hildreth, Abdalla R. Nassar, Kevin R. Chasse, and Timothy W. Simpson. Dissolvable metal supports for 3D direct metal printing. *3D Printing and Additive Manufacturing*, 3(2): 90–97, 2016. DOI: 10.1089/3dp.2016.0013. 68

Scott J. Hollister. Porous scaffold design for tissue engineering. *Nature Materials*, 4(7):518–524, 2005. DOI: 10.1038/nmat1683. 73

Tim Van Hook. Real-time shaded NC milling display. In *Proc. of SIGGRAPH*, pages 15–20, 1986. DOI: 10.1145/15886.15887. 69, 77, 78

R. L. Hope, R. N. Roth, and P. A. Jacobs. Adaptive slicing with sloping layer surfaces. *Rapid Prototyping Journal*, 3(3):89–98, 1997. DOI: 10.1108/13552549710185662. 76

Samuel Hornus and Sylvain Lefebvre. Iterative carving for self-supporting 3D printed cavities. *Research Report RR-9083*, Inria Nancy—Grand Est., July 2017. https://hal.inria.fr/hal-01570330 72

Samuel Hornus, Sylvain Lefebvre, Jérémie Dumas, and Frédéric Claux. Tight printable enclosures and support structures for additive manufacturing. In *EG Workshop on Graphics for Digital Fabrication*, 2016. 72

Kailun Hu, Shuo Jin, and Charlie C. L. Wang. Support slimming for single material based additive manufacturing. *Computer-Aided Design*, 65:1–10, 2015. DOI: 10.1016/j.cad.2015.03.001. 63

Ruizhen Hu, Honghua Li, Hao Zhang, and Daniel Cohen-Or. Approximate pyramidal shape decomposition. *ACM Transactions on Graphics*, 33(6):213:1–213:12, 2014. DOI: 10.1145/2661229.2661244. 59

B. Huang and S. Singamneni. Alternate slicing and deposition strategies for fused deposition modelling of light curved parts. *Journal of Achievements in Materials and Manufacturing*, 55 (2):511–517, 2012. 80

Pu Huang, Charlie C. L. Wang, and Yong Chen. Intersection-free and topologically faithful slicing of implicit solid. *Journal of Computing and Information Science in Engineering*, 13(2): 021009, 2013. DOI: 10.1115/1.4024067. 56, 78

Pu Huang, Charlie C.L. Wang, and Yong Chen. Algorithms for layered manufacturing in image space. *ASME Advances in Computers and Information in Engineering Research*, 1:377–410, 2014a. DOI: 10.1115/1.860328_ch15. 66, 67, 78

Qiang Huang, Hadis Nouri, Kai Xu, Yong Chen, Sobambo Sosina, and Tirthankar Dasgupta. Predictive modeling of geometric deviations of 3D printed products—a unified modeling approach for cylindrical and polygon shapes. In *IEEE International Conference on Automation Science and Engineering*, pages 25–30, 2014b. DOI: 10.1109/coase.2014.6899299. 14

Tingting Huang, Shanggang Wang, and Ketai He. Quality control for fused deposition modeling based additive manufacturing: Current research and future trends. In

1st International Conference on Reliability Systems Engineering, pages 1–6, 2015. DOI: 10.1109/icrse.2015.7366500. 14

Xiaomao Huang, Chunsheng Ye, Jianhua Mo, and Haitao Liu. Slice data based support generation algorithm for fused deposition modeling. *Tsinghua Science and Technology*, 14(S1): 223–228, 2009a. DOI: 10.1016/s1007-0214(09)70096-3. 66

Xiaomao Huang, Chunsheng Ye, Siyu Wu, Kaibo Guo, and Jianhua Mo. Sloping wall structure support generation for fused deposition modeling. *International Journal of Advanced Manufacturing Technology*, 42(11-12):1074–1081, 2009b. DOI: 10.1007/s00170-008-1675-2. 66

Yijiang Huang, Juyong Zhang, Xin Hu, Guoxian Song, Zhongyuan Liu, Lei Yu, and Ligang Liu. Framefab: Robotic fabrication of frame shapes. *ACM Transactions on Graphics*, 35(6), 2016. DOI: 10.1145/2980179.2982401. 80

Kin Chuen Hui. Solid sweeping in image space—application in NC simulation. *The Visual Computer*, 10(6):306–316, 1994. DOI: 10.1007/bf01900825. 69

Matthias B. Hullin, Ivo Ihrke, Wolfgang Heidrich, Tim Weyrich, Gerwin Damberg, and Martin Fuchs. Computational fabrication and display of material appearance. *Eurographics State-of-the-Art Reports (STAR)*, page 17, May 2013. 17, 19

Junghoon Hur and Kunwoo Lee. The development of a CAD environment to determine the preferred build-up direction for layered manufacturing. *International Journal of Advanced Manufacturing Technology*, 14(4):247–254, 1998. DOI: 10.1007/bf01199879. 60, 63

T. Ingrassia, Vincenzo Nigrelli, V. Ricotta, and C. Tartamella. Process parameters influence in additive manufacturing. In *Advance on Mechanics, Design Engineering and Manufacturing*, pages 261–270, Springer, 2017. DOI: 10.1007/978-3-319-45781-9_27. 85

ISO. ISO 4287: 1997: Geometrical product specifications (GPS)—surface texture: Profile method—terms, definitions and surface texture parameters, 1997. 13

Vikram Iyer, Justin Chan, and Shyamnath Gollakota. 3D printing wireless connected objects. *ACM Transactions on Graphics (TOG)*, 36(6):242, 2017. DOI: 10.1145/3130800.3130822. 90

T. R. Jackson, H-Liu, N. M. Patrikalakis, E. M. Sachs, and M. J. Cima. Modeling and designing functionally graded material components for fabrication with local composition control. *Materials and Design*, 20(2–3):63–75, 1999. DOI: 10.1016/s0261-3069(99)00011-4. 73

Alec Jacobson, Ladislav Kavan, and Olga Sorkine-Hornung. Robust inside-outside segmentation using generalized winding numbers. *ACM Transactions on Graphics*, 32(4):33:1–33:12, 2013. DOI: 10.1145/2461912.2461916. 56

Alec Jacobson, Zhigang Deng, Ladislav Kavan, and J. P. Lewis. Skinning: Real-time shape deformation. In *ACM SIGGRAPH Courses*, 2014. DOI: 10.1145/2659467.2675048. 30

Ron Jamieson and Herbert Hacker. Direct slicing of CAD models for rapid prototyping. *Rapid Prototyping Journal*, 1(2):4–12, 1995. DOI: 10.1108/13552549510086826. 79

Jamasp Jhabvala, Eric Boillat, Cédric André, and Rémy Glardon. An innovative method to build support structures with a pulsed laser in the selective laser melting process. *International Journal of Advanced Manufacturing Technology*, 59(1-4):137–142, 2012. DOI: 10.1007/s00170-011-3470-8. 68

Yu An Jin, Yong He, and Jian Zhong Fu. An adaptive tool path generation for fused deposition modeling. In *Advanced Materials Research*, Vol. 819, pages 7–12, Trans. Tech. Publications, 2013. DOI: 10.4028/www.scientific.net/amr.819.7. 71

Yu-an Jin, Yong He, Jian-zhong Fu, Wen-feng Gan, and Zhi-wei Lin. Optimization of tool-path generation for material extrusion-based additive manufacturing technology. *Additive Manufacturing*, 1:32–47, 2014. DOI: 10.1016/j.addma.2014.08.004. 84

Yuan Jin, S. Joe Qin, and Qiang Huang. Out-of-plane geometric error prediction for additive manufacturing. In *IEEE International Conference on Automation Science and Engineering*, pages 918–923, 2015. DOI: 10.1109/coase.2015.7294216. 14

Yuan Jin, Jianke Du, Zhiyong Ma, Anbang Liu, and Yong He. An optimization approach for path planning of high-quality and uniform additive manufacturing. *The International Journal of Advanced Manufacturing Technology*, pages 1–12, 2017. DOI: 10.1007/s00170-017-0207-3. 83

Tao Ju. Robust repair of polygonal models. *ACM Transactions on Graphics*, 23(3):888–895, 2004. DOI: 10.1145/1015706.1015815. 56

Tao Ju. Fixing geometric errors on polygonal models: A survey. *Computer Science and Technology*, 24(1):19–29, 2009. DOI: 10.1007/s11390-009-9206-7. 56

Surya R. Kalidindi. Data science and cyber infrastructure: Critical enablers for accelerated development of hierarchical materials. *International Materials Reviews*, 60(3):150–168, 2015. DOI: 10.1179/1743280414y.0000000043. 24

Ju-Hsien Kao and Fritz B. Prinz. Optimal motion planning for deposition in layered manufacturing. In *Proc. of DETC*, Vol. 98, pages 13–16, 1998. 71, 84

Rahul Khardekar and Sara McMains. Fast layered manufacturing support volume computation on GPUs. In *ASME International Design Engineering Technical Conferences and Computers and Information in Engineering Conference*, pages 993–1002, 2006. DOI: 10.1115/detc2006-99666. 61

Lily Kharevych, Patrick Mullen, Houman Owhadi, and Mathieu Desbrun. Numerical coarsening of inhomogeneous elastic materials. *ACM Transactions on Graphics*, 28(3):51:1–51:8, July 2009. DOI: 10.1145/1531326.1531357. 24

Zhong Xun Khoo, Joanne Ee Mei Teoh, Yong Liu, Chee Kai Chua, Shoufeng Yang, Jia An, Kah Fai Leong, and Wai Yee Yeong. 3D printing of smart materials: A review on recent progresses in 4D printing. *Virtual and Physical Prototyping*, 10(3):103–122, 2015. DOI: 10.1080/17452759.2015.1097054. 90

Martin Kilian, Aron Monszpart, and Niloy J. Mitra. String actuated curved folded surfaces. *ACM Transactions on Graphics*, 36(3):25:1–25:13, May 2017. DOI: 10.1145/3072959.3126802. 41

Bo H. Kim and Byoung K. Choi. Machining efficiency comparison direction-parallel tool path with contour-parallel tool path. *Computer-Aided Design*, 34(2):89–95, 2002. DOI: 10.1016/s0010-4485(00)00139-1. 71

C. Kirschman and C. Jara-Almonte. A parallel slicing algorithm for solid freeform fabrication processes. In *Solid Freeform Fabrication Proc.*, pages 26–33, 1992. 77

C. F. Kirschman, C. C. Jara-Almonte, A. Bagchi, R. L. Dooley, and A. A. Ogale. Computer aided design of support structures for stereolithographic components. In *Proc. of the ASME Computers in Engineering Conference*, pages 443–451, 1991. 66

Mina Konaković, Keenan Crane, Bailin Deng, Sofien Bouaziz, Daniel Piker, and Mark Pauly. Beyond developable: Computational design and fabrication with auxetic materials. *ACM Transactions on Graphics*, 35(4):89:1–89:11, July 2016. DOI: 10.1145/2897824.2925944. 40, 41, 50, 53

Bongjin Koo, Wilmot Li, JiaXian Yao, Maneesh Agrawala, and Niloy J. Mitra. Creating works-like prototypes of mechanical objects. *ACM Transactions on Graphics (Special issue of SIG-GRAPH Asia)*, 2014. DOI: 10.1145/2661229.2661289. 36, 47, 52, 53

X. Y. Kou and S. T. Tan. Heterogeneous object modeling: A review. *Computer-Aided Design*, 39(4):284–301, 2007. DOI: 10.1016/j.cad.2006.12.007. 73

Yuki Koyama, Shinjiro Sueda, Emma Steinhardt, Takeo Igarashi, Ariel Shamir, and Wojciech Matusik. Autoconnect: Computational design of 3D-printable connectors. *ACM Transactions on Graphics*, 34(6):231:1–231:11, October 2015. DOI: 10.1145/2816795.2818060. 37, 51, 52

Tim Kuipers, Eugeni Doubrovski, and Jouke Verlinden. 3D hatching: Linear halftoning for dual extrusion fused deposition modeling. In *Proc. of the 1st Annual ACM Symposium on Computational Fabrication, (SCF'17)*, pages 2:1–2:7, 2017. DOI: 10.1145/3083157.3083163. 81

Prashant Kulkarni and Debasish Dutta. An accurate slicing procedure for layered manufacturing. *Computer-Aided Design*, 28(9):683–697, 1996. DOI: 10.1016/0010-4485(95)00083-6. 76

Gurunathan Saravana Kumar, Ponnusamy Pandithevan, and Appa Rao Ambatti. Fractal raster tool paths for layered manufacturing of porous objects. *Virtual and Physical Prototyping*, 4(2): 91–104, 2009. DOI: 10.1080/17452750802688215. 72

Vinod Kumar and Debasish Dutta. An approach to modeling and representation of heterogeneous objects. *Journal of Mechanical Design*, 120(4):659, 1998. DOI: 10.1115/1.2829329. 80

Vinod Kumar, Prashant Kulkarni, and Debasish Dutta. Adaptive slicing of heterogeneous solid models for layered manufacturing. In *Proc. of the ASME Design Technical Conferences*, 1998. DOI: 10.1106/7r2c-whgu-m1jb-9h9b. 80

Amirali Lalehpour, Conner Janeteas, and Ahmad Barari. Surface roughness of FDM parts after post-processing with acetone vapor bath smoothing process. *The International Journal of Advanced Manufacturing Technology*, pages 1–16, 2017. DOI: 10.1007/s00170-017-1165-5. 13

Po-Ting Lan, Shuo-Yan Chou, Lin-Lin Chen, and Douglas Gemmill. Determining fabrication orientations for rapid prototyping with stereolithography apparatus. *Computer-Aided Design*, 29(1):53–62, 1997. DOI: 10.1016/s0010-4485(96)00049-8. 60

Timothy Langlois, Ariel Shamir, Daniel Dror, Wojciech Matusik, and David I.W. Levin. Stochastic structural analysis for context-aware design and fabrication. *ACM Transactions on Graphics*, 35(6), 2016. DOI: 10.1145/2980179.2982436. 14, 22

Gierad Laput, Xiang A. Chen, and Chris Harrison. 3D printed hair: Fused deposition modeling of soft strands, fibers, and bristles. In *Proc. of UIST*, pages 593–597, 2015. DOI: 10.1145/2807442.2807484. 80

Manfred Lau, Akira Ohgawara, Jun Mitani, and Takeo Igarashi. Converting 3D furniture models to fabricatable parts and connectors. In *ACM SIGGRAPH Papers*, pages 85:1–85:6, 2011. DOI: 10.1145/1964921.1964980. 36, 47, 51, 53

Jusung Lee and Kunwoo Lee. Block-based inner support structure generation algorithm for 3D printing using fused deposition modeling. *International Journal of Advanced Manufacturing Technology*, 2016. DOI: 10.1007/s00170-016-9239-3. 72

Taeseok Lee, Jusung Lee, and Kunwoo Lee. Extended block based infill generation. *The International Journal of Advanced Manufacturing Technology*, 93(1):1415–1430, 2017. DOI: 10.1007/s00170-017-0572-y. 72

Sylvain Lefebvre. IceSL : A GPU accelerated modeler and slicer. In *18th European Forum on Additive Manufacturing*, 2013. 78

Sylvain Lefebvre. 3D infilling: faster, stronger, simpler, 2015. `http://sylefeb.blogspot.fr/2015/07/3dprint-3d-infilling-faster-stronger.html` 72, 73

Sylvain Lefebvre and Salim Perchy. IceSL. `http://www.loria.fr/~slefebvr/icesl/`, 2013. 56

Sylvain Lefebvre, Samuel Hornus, and Anass Lasram. Per-pixel lists for single pass a-buffer. In *GPU Pro5*, pages 3–23, Informa UK Ltd., 2014. DOI: 10.1201/b16721-3. 77

Yuen-Shan Leung and C. C. L. Wang. Conservative sampling of solids in image space. *IEEE Computer Graphics and Applications*, 33(1):32–43, 2013. DOI: 10.1109/mcg.2013.2. 77

Dawei Li, Ning Dai, Xiaotong Jiang, and Xiaosheng Chen. Interior structural optimization based on the density-variable shape modeling of 3D printed objects. *International Journal of Advanced Manufacturing Technology*, 83(9):1627–1635, 2015a. DOI: 10.1007/s00170-015-7704-z. 14, 74

Dingzeyu Li, David I. W. Levin, Wojciech Matusik, and Changxi Zheng. Acoustic voxels: Computational optimization of modular acoustic filters. *ACM Transactions on Graphics*, 35 (4):88, 2016. DOI: 10.1145/2897824.2925960. 35, 47, 52

Honghua Li, Ruizhen Hu, Ibraheem Alhashim, and Hao Zhang. Foldabilizing furniture. *ACM Transactions on Graphics, (Proc. of SIGGRAPH)*, 34(4), 2015b. DOI: 10.1145/2766912. 36

Wei Li and Sara McMains. A sweep and translate algorithm for computing voxelized 3D Minkowski sums on the GPU. *Computer-Aided Design*, 46(0):90–100, 2014. DOI: 10.1016/j.cad.2013.08.021. 69

Xian-Ying Li, Chao-Hui Shen, Shi-Sheng Huang, Tao Ju, and Shi-Min Hu. Popup: automatic paper architectures from 3D models. *ACM Transactions on Graphics*, 29(4):111:1–9, 2010. DOI: 10.1145/1833351.1778848. 41

Xian-Ying Li, Tao Ju, Yan Gu, and Shi-Min Hu. A geometric study of v-style pop-ups: Theories and algorithms. In *ACM SIGGRAPH Papers*, pages 98:1–98:10, 2011. DOI: 10.1145/1964921.1964993. 41

Jyh-Ming Lien. Covering Minkowski sum boundary using points with applications. *Computer Aided Geometric Design*, 25(8):652–666, 2008. DOI: 10.1016/j.cagd.2008.06.006. 69

G. H. Liu, Y. S. Wong, Y. F. Zhang, and Han Tong Loh. Error-based segmentation of cloud data for direct rapid prototyping. *Computer-Aided Design*, 35(7):633–645, 2003. DOI: 10.1016/s0010-4485(02)00087-8. 79

Shengjun Liu and Charlie C. L. Wang. Fast intersection-free offset surface generation from freeform models with triangular meshes. *IEEE Transactions on Automation Science Engineering*, 8(2):347–360, 2011. DOI: 10.1109/tase.2010.2066563. 69

Yang Liu, Helmut Pottmann, Johannes Wallner, Yong-Liang Yang, and Wenping Wang. Geometric modeling with conical meshes and developable surfaces. *ACM Transactions on Graphics*, 25(3):681–689, July 2006. DOI: 10.1145/1141911.1141941. 43, 49, 51

Marco Livesu, Stefano Ellero, Jonás Martìnez, Sylvain Lefebvre, and Marco Attene. From 3D models to 3D prints: An overview of the processing pipeline. *Computer Graphics Forum (Eurographics STAR)*, 36(2), 2017. DOI: 10.1111/cgf.13147. 2, 6, 10, 12, 65, 70, 75, 83

Kui-Yip Lo, Chi-Wing Fu, and Hongwei Li. 3D polyomino puzzle. *ACM Transactions on Graphics*, 28(5):157:1–157:8, December 2009. DOI: 10.1145/1618452.1618503. 38, 48, 50, 53

Lin Lu, Andrei Sharf, Haisen Zhao, Yuan Wei, Qingnan Fan, Xuelin Chen, Yann Savoye, Changhe Tu, Daniel Cohen-Or, and Baoquan Chen. Build-to-last: Strength to weight 3D printed objects. *ACM Transactions on Graphics*, 33(4):97, 2014. DOI: 10.1145/2601097.2601168. 11, 23, 47, 50, 52, 73

Linjie Luo, Ilya Baran, Szymon Rusinkiewicz, and Wojciech Matusik. Chopper: Partitioning models into 3D-printable parts. *ACM Transactions on Graphics*, 31(6):129, 2012. DOI: 10.1145/2366145.2366148. 56, 57, 58

Li-Ke Ma, Yizhonc Zhang, Yang Liu, Kun Zhou, and Xin Tong. Computational design and fabrication of soft pneumatic objects with desired deformations. *ACM Transactions on Graphics*, 36(6):239:1–239:12, November 2017. DOI: 10.1145/3130800.3130850. 28

Weiyin Ma, Wing-Chung But, and Peiren He. NURBS-based adaptive slicing for efficient rapid prototyping. *Computer-Aided Design*, 36(13):1309–1325, 2004. DOI: 10.1016/j.cad.2004.02.001. 79

Anant Madabhushi, Michael D. Feldman, Dimitris N. Metaxas, John Tomaszeweski, and Deborah Chute. Automated detection of prostatic adenocarcinoma from high-resolution ex vivo MRI. *IEEE Transactions on Medical Imaging*, 24(12):1611–1625, 2005. DOI: 10.1109/tmi.2005.859208. 49

Stanislav S. Makhanov and Weerachai Anotaipaiboon. *Advanced Numerical Methods to Optimize Cutting Operations of Five Axis Milling Machines*. Springer Science and Business Media, 2007. DOI: 10.1007/978-3-540-71121-6. 71, 83

Luigi Malomo, Nico Pietroni, Bernd Bickel, and Paolo Cignoni. Flexmolds: Automatic design of flexible shells for molding. *ACM Transactions on Graphics*, 35(6):223:1–223:12, November 2016. DOI: 10.1145/2980179.2982397. 27

Ka Mani, Prashant Kulkarni, and Debasish Dutta. Region-based adaptive slicing. *Computer-Aided Design*, 31(5):317–333, 1999. DOI: 10.1016/s0010-4485(99)00033-0. 76

Tobias Martin, Nobuyuki Umetani, and Bernd Bickel. Omniad: Data-driven omni-directional aerodynamics. *ACM Transactions on Graphics*, 34(4):113:1–113:12, July 2015. DOI: 10.1145/2766919. 34, 50, 52

Jonàs Martínez, Jérémie Dumas, Sylvain Lefebvre, and Li-Yi Wei. Structure and appearance optimization for controllable shape design. *ACM Transactions on Graphics*, 34(6):229:1–229:11, October 2015. DOI: 10.1145/2816795.2818101. 19, 52

Jonàs Martínez, Samuel Hornus, Frédéric Claux, and Sylvain Lefebvre. Chained segment offsetting for ray-based solid representations. *Computers and Graphics*, 46:36–47, 2015. DOI: 10.1016/j.cag.2014.09.017. 69

Jonàs Martínez, Jérémie Dumas, and Sylvain Lefebvre. Procedural Voronoi foams for additive manufacturing. *ACM Transactions on Graphics*, 35(4):44:1–44:12, 2016. DOI: 10.1145/2897824.2925922. 27, 47, 52, 74

Jonàs Martínez, Haichuan Song, Jérémie Dumas, and Sylvain Lefebvre. Orthotropic k-nearest foams for additive manufacturing. *ACM Transactions on Graphics*, 36(4): 121:1–121:12, July 2017. https://hal.archives-ouvertes.fr/hal-01577859 DOI: 10.1145/3072959.3073638. 27, 74

S. H. Masood, W. Rattanawong, and P. Iovenitti. Part build orientations based on volumetric error in fused deposition modelling. *International Journal of Advanced Manufacturing Technology*, 16(3):162–168, 2000. DOI: 10.1007/s001700050022. 12, 61

S. H. Masood, W. Rattanawong, and P. Iovenitti. A generic algorithm for a best part orientation system for complex parts in rapid prototyping. *Journal of Materials Processing Technology*, 139 (1):110–116, 2003. DOI: 10.1016/s0924-0136(03)00190-0. 13, 61

Fady Massarwi, Craig Gotsman, and Gershon Elber. Papercraft models using generalized cylinders. In *Computer Graphics and Applications, (PG'07), 15th Pacific Conference on*, pages 148–157, IEEE, 2007. DOI: 10.1109/pg.2007.16. 41

Marilena Maule, João Luiz Dihl Comba, Rafael P. Torchelsen, and Rui Bastos. A survey of raster-based transparency techniques. *Computers and Graphics*, 35(6):1023–1034, 2011. DOI: 10.1016/j.cag.2011.07.006. 77

James McCrae, Nobuyuki Umetani, and Karan Singh. Flatfitfab: Interactive modeling with planar sections. In *Proc. of the 27th Annual ACM Symposium on User Interface Software and Technology, (UIST'14)*, pages 13–22, New York, 2014. DOI: 10.1145/2642918.2647388. 42, 43

Sara McMains and Carlo Séquin. A coherent sweep plane slicer for layered manufacturing. In *Proc. of the 5th ACM Symposium on Solid Modeling and Applications*, 1999. DOI: 10.1145/304012.304042. 77

Sara McMains, Jordan Smith, Jianlin Wang, and Carlo Séquin. Layered manufacturing of thin-walled parts. In *ASME Design Engineering Technical Conference*, 2000. 69, 70, 71

Mcor. Mcor technologies, 2005. http://mcortechnologies.com/ 4, 81

Asla Medeiros e Sá, Vinícius Moreira Mello, Karina Rodriguez Echavarria, and Derek Covill. Adaptive voids. *Visual Computer*, 31(6-8):799–808, 2015. DOI: 10.1007/s00371-015-1109-8. 72

Vittorio Megaro, Bernhard Thomaszewski, Maurizio Nitti, Otmar Hilliges, Markus Gross, and Stelian Coros. Interactive design of 3D-printable robotic creatures. *ACM Transactions on Graphics*, 34(6):216:1–216:9, October 2015. DOI: 10.1145/2816795.2818137. 32, 52

Vittorio Megaro, Jonas Zehnder, Moritz Bächer, Stelian Coros, Markus Gross, and Bernhard Thomaszewski. A computational design tool for compliant mechanisms. *ACM Transactions on Graphics*, 36(4):82:1–82:12, July 2017. DOI: 10.1145/3072959.3073636. 30

Microsoft and NetFABB. 3D model repair service. https://modelrepair.azurewebsites. net/, 2013. 56

Eder Miguel, Mathias Lepoutre, and Bernd Bickel. Computational design of stable planar-rod structures. *ACM Transactions on Graphics*, 35(4):86:1–86:11, July 2016. DOI: 10.1145/2897824.2925978. 39, 40

Masaaki Miki, Takeo Igarashi, and Philippe Block. Parametric self-supporting surfaces via direct computation of airy stress functions. *ACM Transactions on Graphics*, 34(4):89:1–89:12, July 2015. DOI: 10.1145/2766888. 50

Jun Mitani and Hiromasa Suzuki. Making papercraft toys from meshes using strip-based approximate unfolding. *ACM Transactions on Graphics (TOG)*, 23(3):259–263, 2004. DOI: 10.1145/1015706.1015711. 41

Mitutoyo. Surface roughness measurement. *Mitutoyo America Corporation*, Buletin n. 1984, 2009. 13

H. D. Morgan, J. A. Cherry, S. Jonnalagadda, D. Ewing, and J. Sienz. Part orientation optimisation for the additive layer manufacture of metal components. *International Journal of Advanced Manufacturing Technology*, 86(5):1679–1687, 2016. DOI: 10.1007/s00170-015-8151-6. 61

Yuki Mori and Takeo Igarashi. Plushie: An interactive design system for plush toys. In *ACM SIGGRAPH Papers*, 2007. DOI: 10.1145/1275808.1276433. 39, 49, 53

Stefanie Mueller, Sangha Im, Serafima Gurevich, Alexander Teibrich, Lisa Pfisterer, François Guimbretière, and Patrick Baudisch. Wireprint: 3D printed previews for fast prototyping. In *Proc. of UIST*, 2014. DOI: 10.1145/2642918.2647359. 4, 48, 80

Przemyslaw Musialski, Thomas Auzinger, Michael Birsak, Michael Wimmer, and Leif Kobbelt. Reduced-order shape optimization using offset surfaces. *ACM Transactions on Graphics*, 34 (4):102, 2015. DOI: 10.1145/2766955. 15, 34, 48, 50

Przemyslaw Musialski, Christian Hafner, Florian Rist, Michael Birsak, Michael Wimmer, and Leif Kobbelt. Non-linear shape optimization using local subspace projections. *ACM Transactions on Graphics*, 35(4):87:1–87:13, July 2016a. DOI: 10.1145/2897824.2925886. 34

Przemyslaw Musialski, Christian Hafner, Florian Rist, Michael Birsak, Michael Wimmer, and Leif Kobbelt. Non-linear shape optimization using local subspace projections. *ACM Transactions on Graphics*, 35(4):87:1–87:13, 2016b. DOI: 10.1145/2897824.2925886. 15

Andrew Nealen, Matthias Müller, Richard Keiser, Eddy Boxerman, and Mark Carlson. Physically based deformable models in computer graphics. *Computer Graphics Forum*, 25(4):809–836, 2006. DOI: 10.1111/j.1467-8659.2006.01000.x. 22

Stephen T. Newman, Zicheng Zhu, Vimal Dhokia, and Alborz Shokrani. Process planning for additive and subtractive manufacturing technologies. *CIRP Annals-Manufacturing Technology*, 64(1):467–470, 2015. DOI: 10.1016/j.cirp.2015.04.109. 87

Griffin Nicoll. Rhombic dodecahedron infill, 2011. http://www.thingiverse.com/thing: 12535 72

Neri Oxman. Variable property rapid prototyping. *Virtual and Physical Prototyping*, 6(1):3–31, 2011. DOI: 10.1080/17452759.2011.558588. 73

Yayue Pan and Yong Chen. Smooth surface fabrication based on controlled meniscus and cure depth in microstereolithography. *Journal of Micro and Nano-Manufacturing*, 3(3):031001, 2015. DOI: 10.1115/1.4030661. 80

Yayue Pan, Xuejin Zhao, Chi Zhou, and Yong Chen. Smooth surface fabrication in mask projection based stereolithography. *Journal of Manufacturing Processes*, 14(4):460–470, 2012. DOI: 10.1016/j.jmapro.2012.09.003. 80

P. M. Pandey, N. Venkata Reddy, and S. G. Dhande. Part deposition orientation studies in layered manufacturing. *Journal of Materials Processing Technology*, 185(1):125–131, 2007. DOI: 10.1016/j.jmatprotec.2006.03.120. 60

Julian Panetta, Qingnan Zhou, Luigi Malomo, Nico Pietroni, Paolo Cignoni, and Denis Zorin. Elastic textures for additive fabrication. *ACM Transactions on Graphics*, 34(4):135:1–135:12, 2015. DOI: 10.1145/2766937. 25, 47, 52, 74

Julian Panetta, Abtin Rahimian, and Denis Zorin. Worst-case stress relief for microstructures. *ACM Transactions on Graphics*, 36(4):122:1–122:16, July 2017. http://doi.acm.org/10.1145/3072959.3073649 DOI: 10.1145/3072959.3073649. 25

Daniele Panozzo, Philippe Block, and Olga Sorkine-Hornung. Designing unreinforced masonry models. *ACM Transactions on Graphics*, 32(4):91:1–91:12, July 2013. DOI: 10.1145/2461912.2461958. 47, 49, 53

Daniele Panozzo, Olga Diamanti, Sylvain Paris, Marco Tarini, Evgeni Sorkine, and Olga Sorkine-Hornung. Texture mapping real-world objects with hydrographics. *Computer Graphics Forum (Proc. of EUROGRAPHICS Symposium on Geometry Processing)*, 34(5):65–75, 2015. DOI: 10.1111/cgf.12697. 17

Marios Papas, Christian Regg, Wojciech Jarosz, Bernd Bickel, Philip Jackson, Wojciech Matusik, Steve Marschner, and Markus Gross. Fabricating translucent materials using continuous pigment mixtures. *ACM Transactions on Graphics*, 32(4):146:1–146:12, July 2013. DOI: 10.1145/2461912.2461974. 18, 52

In Baek Park, Young Myoung Ha, and Seok Hee Lee. Dithering method for improving the surface quality of a microstructure in projection microstereolithography. *International Journal of Advanced Manufacturing Technology*, 52(5-8):545–553, 2011. 80

Sang C. Park. Hollowing objects with uniform wall thickness. *Computer-Aided Design*, 37(4):451–460, 2005. DOI: 10.1016/j.cad.2004.08.001. 70

Seok-Min Park, Richard H. Crawford, and Joseph J. Beaman. Volumetric multi-texturing for functionally gradient material representation. In *ACM Symposium on Solid Modeling and Applications*, 2001. DOI: 10.1145/376957.376982. 74, 80

Alexander Pasko, Oleg Fryazinov, Turlif Vilbrandt, Pierre-Alain Fayolle, and Valery Adzhiev. Procedural function-based modelling of volumetric microstructures. *Graphical Models*, 73(5):165–181, 2011. DOI: 10.1016/j.gmod.2011.03.001. 74

Darko Pavić and Leif Kobbelt. High-resolution volumetric computation of offset surfaces with feature preservation. *Computer Graphics Forum*, 27(2):165–174, 2008. DOI: 10.1111/j.1467-8659.2008.01113.x. 69

Huaishu Peng, François Guimbretière, James McCann, and Scott Hudson. A 3D printer for interactive electromagnetic devices. In *Proc. of UIST*, pages 553–562, 2016. DOI: 10.1145/2984511.2984523. 90

Thiago Pereira, Szymon Rusinkiewicz, and Wojciech Matusik. Computational light routing: 3D printed optical fibers for sensing and display. *ACM Transactions on Graphics*, 33(3):24, 2014. DOI: 10.1145/2602140. 18, 47, 50, 52, 53

Thiago Pereira, Carolina L. A. Paes Leme, Steve Marschner, and Szymon Rusinkiewicz. Printing anisotropic appearance with magnetic flakes. *ACM Transactions on Graphics*, 36(4):123, July 2017. DOI: 10.1145/3072959.3073701. 18

Jesús Pérez, Bernhard Thomaszewski, Stelian Coros, Bernd Bickel, José A. Canabal, Robert Sumner, and Miguel A. Otaduy. Design and fabrication of flexible rod meshes. *ACM Transactions on Graphics*, 34(4):138:1–138:12, July 2015. DOI: 10.1145/2766998. 28, 48, 49, 52

D. T. Pham, S. S. Dimov, and R. S. Gault. Part orientation in stereolithography. *International Journal of Advanced Manufacturing Technology*, 15(9):674–682, 1999. DOI: 10.1007/s001700050118. 10, 63

Amar M. Phatak and S. S. Pande. Optimum part orientation in rapid prototyping using genetic algorithm. *Journal of Manufacturing Systems*, 31(4):395–402, 2012. DOI: 10.1016/j.jmsy.2012.07.001. 63

Nico Pietroni, Davide Tonelli, Enrico Puppo, Maurizio Froli, Roberto Scopigno, and Paolo Cignoni. Voronoi grid-shell structures. *CoRR*, abs/1408.6591, 2014. http://arxiv.org/abs/1408.6591 44, 47, 49, 53

Nico Pietroni, Marco Tarini, Amir Vaxman, Daniele Panozzo, and Paolo Cignoni. Position-based tensegrity design. *ACM Transactions on Graphics*, 36(6):172:1–172:14, November 2017. DOI: 10.1145/3130800.3130809. 29

PolyJet. Polyjet technology, 1998. https://en.wikipedia.org/wiki/Objet_Geometries 3

Helmut Pottmann. Architectural geometry and fabrication-aware design. *Nexus Network Journal*, 15(2):195–208, 2013. DOI: 10.1007/s00004-013-0149-5. 43

Romain Prévost, Emily Whiting, Sylvain Lefebvre, and Olga Sorkine-Hornung. Make it stand: Balancing shapes for 3D fabrication. *ACM Transactions on Graphics*, 32(4):81:1–81:10, 2013. DOI: 10.1145/2461912.2461957. 14, 33, 47

Romain Prévost, Moritz Bächer, Wojciech Jarosz, and Olga Sorkine-Hornung. Balancing 3D models with movable masses. In *Proc. of the Vision, Modeling and Visualization Workshop*, 2016. 15

William R. Priedeman Jr. and Andrea L. Brosch. Soluble material and process for three-dimensional modeling, September 2004. *U.S. Patent 6790403*. 68

Christopher G. Provatidis. Analysis of box-like structures using 3D coons' interpolation. *Communications in Numerical Methods in Engineering*, 21(8):443–456, 2005. DOI: 10.1002/cnm.762. 51

Di Qi, Long Zeng, and Matthew M. F. Yuen. Robust slicing procedure based on surfel-grid. *Computer Aided Design Applications*, 10(6):965–981, 2013. DOI: 10.3722/cadaps.2013.965-981. 78

Yanjie Qiu, Xionghui Zhou, and Xiaoping Qian. Direct slicing of cloud data with guaranteed topology for rapid prototyping. *International Journal of Advanced Manufacturing Technology*, 53(1):255–265, 2011. DOI: 10.1007/s00170-010-2829-6. 79

Xiuzhi Qu and Brent Stucker. A 3D surface offset method for STL format models. *Rapid Prototyping Journal*, 9(3):133–141, 2003. DOI: 10.1108/13552540310477436. 68

Zhenzhen Quan, Jonghwan Suhr, Jianyong Yu, Xiaohong Qin, Chase Cotton, Mark Mirotznik, and Tsu-Wei Chou. Printing direction dependence of mechanical behavior of additively manufactured 3D preforms and composites. *Composite Structures*, 2017. DOI: 10.1016/j.compstruct.2017.10.055. 62

Jordan R. Raney and Jennifer A. Lewis. Printing mesoscale architectures. *MRS Bulletin*, 40 (11):943–950, 2015. DOI: 10.1557/mrs.2015.235. 74, 90

Philip E. Reeves and Richard C. Cobb. Reducing the surface deviation of stereolithography using in-process techniques. *Rapid Prototyping Journal*, 3(1):20–31, 1997. DOI: 10.1108/13552549710169255. 13

Tim Reiner and Sylvain Lefebvre. Interactive modeling of support-free shapes for fabrication. In *Eurographics, Short Papers*, 2016. 63

Tim Reiner, Nathan Carr, Radomír Měch, Ondřej Šťava, Carsten Dachsbacher, and Gavin Miller. Dual-color mixing for fused deposition modeling printers. *Computer Graphics Forum*, 33(2):479–486, 2014. DOI: 10.1111/cgf.12319. 18, 48, 52, 81, 82

David Thompson Richard and Richard H. Crawford. Optimizing part quality with orientation. The University of Texas at Austin, 1995. 60

Xavier Rolland-Neviere, Gwenael Doerr, and Pierre Alliez. Robust diameter-based thickness estimation of 3D objects. *Graphical Models*, 75:279–296, 2013. DOI: 10.1016/j.gmod.2013.06.001. 55

David W. Rosen. Computer-aided design for additive manufacturing of cellular structures. *Computer Aided Design Applications*, 4(5), 2007. DOI: 10.1080/16864360.2007.10738493. 73, 79

Jaroslaw R. Rossignac and Aristides A. G. Requicha. Offsetting operations in solid modelling. *Compututer Aided Geometric Design*, 3(2):129–148, 1986. DOI: 10.1016/0167-8396(86)90017-8. 68

Emmanuel Sabourin, Scott A. Houser, and Jan Helge Bøhn. Adaptive slicing using stepwise uniform refinement. *Rapid Prototyping Journal*, 2(4):20–26, 1996. DOI: 10.1108/13552549610153370. 76

Emmanuel Sabourin, Scott A. Houser, and Jan Helge Bøhn. Accurate exterior, fast interior layered manufacturing. *Rapid Prototyping Journal*, 3(2):44–52, 1997. DOI: 10.1108/13552549710176662. 76

Valkyrie Savage, Xiaohan Zhang, and Björn Hartmann. Midas: Fabricating custom capacitive touch sensors to prototype interactive objects. In *Proc. of UIST*, pages 579–588, 2012. DOI: 10.1145/2380116.2380189. 90

Valkyrie Savage, Colin Chang, and Björn Hartmann. Sauron: Embedded single-camera sensing of printed physical user interfaces. In *Proc. of UIST*, pages 447–456, ACM, 2013. DOI: 10.1145/2501988.2501992. 90

Ryan Schmidt and Nobuyuki Umetani. Branching support structures for 3D printing. In *ACM SIGGRAPH Studio*, page 9, 2014. DOI: 10.1145/2619195.2656293. 66

Christian Schüller, Daniele Panozzo, Anselm Grundhöfer, Henning Zimmer, Evgeni Sorkine, and Olga Sorkine-Hornung. Computational thermoforming. *Transactions on Graphics (Proc. of ACM SIGGRAPH)*, 35(4), 2016. DOI: 10.1145/2897824.2925914. 17

Adriana Schulz, Ariel Shamir, David I. W. Levin, Pitchaya Sitthi-Amorn, and Wojciech Matusik. Design and fabrication by example. *ACM Transactions on Graphics (Proc. SIGGRAPH)*, 33(4), 2014. DOI: 10.1145/2601097.2601127. 36, 51

Adriana Schulz, Jie Xu, Bo Zhu, Changxi Zheng, Eitan Grinspun, and Wojciech Matusik. Interactive design space exploration and optimization for cad models. *ACM Transactions on Graphics*, 36(4):157:1–157:14, July 2017. http://doi.acm.org/10.1145/3072959.3073688 DOI: 10.1145/3072959.3073688. 33

Christian Schumacher, Bernd Bickel, Jan Rys, Steve Marschner, Chiara Daraio, and Markus Gross. Microstructures to control elasticity in 3D printing. *ACM Transactions on Graphics*, 34(4):136:1–136:13, 2015. DOI: 10.1145/2766926. 25, 47, 52, 53, 74

Yuliy Schwartzburg and Mark Pauly. Fabrication-aware design with intersecting planar pieces. *Computer Graphics Forum (Proc. of Eurographics)*, 32(2):317–326, 2013. DOI: 10.1111/cgf.12051. 42

Thomas W. Sederberg, Jianmin Zheng, Almaz Bakenov, and Ahmad Nasri. T-splines and t-nurccs. *ACM Transactions on Graphics*, 22(3):477–484, July 2003. DOI: 10.1145/882262.882295. 51

Carlo H. Séquin. Prototyping dissection puzzles with layered manufacturing. In *Fabrication and Sculpture Track, Shape Modeling International*, 2012. 38

Jonathan Shade, Steven Gortler, Li-wei He, and Richard Szeliski. Layered depth images. In *Proc. of SIGGRAPH*, pages 231–242, 1998. DOI: 10.1145/280814.280882. 77

Hayong Shin, Seyoun Park, and Eonjin Park. Direct slicing of a point set model for rapid prototyping. *Computer Aided Design Applications*, 1(1-4):109–115, 2004. DOI: 10.1080/16864360.2004.10738249. 79

Ki-Hoon Shin, Harshad Natu, Debasish Dutta, and Jyotirmoy Mazumder. A method for the design and fabrication of heterogeneous objects. *Materials and Design*, 24(5):339–353, 2003. DOI: 10.1016/s0261-3069(03)00060-8. 80

Maria Shugrina, Ariel Shamir, and Wojciech Matusik. Fab forms: Customizable objects for fabrication with validity and geometry caching. *ACM Transactions on Graphics*, 34(4), 2015. DOI: 10.1145/2766994. 33, 51, 52

Ole Sigmund. Tailoring materials with prescribed elastic properties. *Mechanics of Materials*, 20 (4):351–368, 1995. DOI: 10.1016/0167-6636(94)00069-7. 73

S. Sikder, A. Barari, and H. A. Kishawy. Global adaptive slicing of NURBS based sculptured surface for minimum texture error in rapid prototyping. *Rapid Prototyping Journal*, 21(6): 649–661, 2015. DOI: 10.1108/rpj-09-2013-0090. 76

Sarat Singamneni, Asimava Roychoudhury, Olaf Diegel, and Bin Huang. Modeling and evaluation of curved layer fused deposition. *Journal of Materials on Processing Technology*, 212(1): 27–35, 2012. DOI: 10.1016/j.jmatprotec.2011.08.001. 80

Mayank Singh and Scott Schaefer. Triangle surfaces with discrete equivalence classes. *ACM Transactions on Graphics*, 29(4):46:1–46:7, July 2010. DOI: 10.1145/1778765.1778783. 44, 49

Pitchaya Sitthi-Amorn, Javier E. Ramos, Yuwang Wangy, Joyce Kwan, Justin Lan, Wenshou Wang, and Wojciech Matusik. MultiFab: A machine vision assisted platform for multi-material 3D printing. *ACM Transactions on Graphics*, 34(4):129:1–129:11, 2015. DOI: 10.1145/2766962. 3, 47, 50

Mélina Skouras, Bernhard Thomaszewski, Bernd Bickel, and Markus Gross. Computational design of rubber balloons. *Computer Graphics Forum (Proc. Eurographics)*, 2012. DOI: 10.1111/j.1467-8659.2012.03064.x. 27, 49, 53

Mélina Skouras, Bernhard Thomaszewski, Stelian Coros, Bernd Bickel, and Markus Gross. Computational design of actuated deformable characters. *ACM Transactions on Graphics*, 32 (4):82:1–82:10, July 2013. DOI: 10.1145/2461912.2461979. 28, 29, 47, 53

Mélina Skouras, Bernhard Thomaszewski, Peter Kaufmann, Akash Garg, Bernd Bickel, Eitan Grinspun, and Markus Gross. Designing inflatable structures. *ACM Transactions on Graphics*, 33(4):63:1–63:10, July 2014. DOI: 10.1145/2601097.2601166. 41, 49, 53

Slicer. Slicer. `http://slic3r.org/`, 2011. 85

Hai-Chuan Song and Sylvain Lefebvre. Colored fused filament fabrication. *CoRR*, abs/1709.09689, 2017. `http://arxiv.org/abs/1709.09689` 81

Peng Song, Chi-Wing Fu, and Daniel Cohen-Or. Recursive interlocking puzzles. *ACM Transactions on Graphics*, 31(6):128:1–128:10, November 2012. DOI: 10.1145/2366145.2366147. 38

Peng Song, Zhongqi Fu, Ligang Liu, and Chi-Wing Fu. Printing 3D objects with interlocking parts. *Computer Aided Geometrics Design*, 35–36:137–148, 2015. DOI: 10.1016/j.cagd.2015.03.020. 58

Peng Song, Bailin Deng, Ziqi Wang, Zhichao Dong, Wei Li, Chi-Wing Fu, and Ligang Liu. CofiFab: Coarse-to-fine fabrication of large 3D objects. *ACM Transactions on Graphics*, 35 (4):45:1–45:11, 2016. DOI: 10.1145/2897824.2925876. 11, 58

Peng Song, Chi-Wing Fu, Yueming Jin, Hongfei Xu, Ligang Liu, Pheng-Ann Heng, and Daniel Cohen-Or. Reconfigurable interlocking furniture. *ACM Transactions on Graphics*, 36(6):174:1–174:14, November 2017. `http://doi.acm.org/10.1145/3130800.3130803` DOI: 10.1145/3130800.3130803. 39

Binil Starly, Alan Lau, Wei Sun, Wing Lau, and Tom Bradbury. Direct slicing of STEP based NURBS models for layered manufacturing. *Computer-Aided Design*, 37(4):387–397, 2005. DOI: 10.1016/j.cad.2004.06.014. 79

Ondrej Stava, Juraj Vanek, Bedrich Benes, Nathan Carr, and Radomír Měch. Stress relief: Improving structural strength of 3D printable objects. *ACM Transactions on Graphics*, 31(4): 48:1–48:11, 2012. DOI: 10.1145/2185520.2335399. 14, 23, 47, 48, 50, 53, 55

John C. Steuben, Athanasios P. Iliopoulos, and John G. Michopoulos. Implicit slicing for functionally tailored additive manufacturing. *Computer-Aided Design*, 77:107–119, 2016. DOI: 10.1016/j.cad.2016.04.003. 72

David Sturman. The state of computer animation. *SIGGRAPH Computer Graphics*, 32(1):57–61, February 1998. DOI: 10.1145/279389.279467. 30

Yong Seok Suh, Michael J. Wozny, et al. Adaptive slicing of solid freeform fabrication processes. In *Solid Freeform Fabrication Symposium*, pages 404–411, 1994. 76

Timothy Sun and Changxi Zheng. Computational design of twisty joints and puzzles. *ACM Transactions on Graphics (Proc. of SIGGRAPH)*, 34(4), August 2015. http://www.cs.columbia.edu/cg/twisty DOI: 10.1145/2766961. 38

Andrea Tagliasacchi, Thomas Delame, Michela Spagnuolo, Nina Amenta, and Alexandru Telea. 3D skeletons: A state-of-the-art report. *Computer Graphics Forum*, 35(2):573–597, 2016. DOI: 10.1111/cgf.12865. 48

Masahito Takezawa, Takuma Imai, Kentaro Shida, and Takashi Maekawa. Fabrication of freeform objects by principal strips. *ACM Transactions on Graphics*, 35(6):225:1–225:12, November 2016. DOI: 10.1145/2980179.2982406. 38

Chengcheng Tang, Pengbo Bo, Johannes Wallner, and Helmut Pottmann. Interactive design of developable surfaces. *ACM Transactions on Graphics*, 35(2):12:1–12:12, January 2016. DOI: 10.1145/2832906. 41

Mohammad Taufik and Prashant K. Jain. Role of build orientation in layered manufacturing: A review. *International Journal of Manufacturing Technology and Management*, 27(1-3):47–73, 2013. DOI: 10.1504/ijmtm.2013.058637. 60

Mohammad Taufik and Prashant Kumar Jain. Volumetric error control in layered manufacturing. In *ASME International Design Engineering Technical Conferences and Computers and Information in Engineering Conference*, 2014. DOI: 10.1115/detc2014-35099. 13

A. Telea and A. Jalba. Voxel-based assessment of printability of 3D shapes. In *Proc. of ISMM*, pages 393–404, Springer, 2011. DOI: 10.1007/978-3-642-21569-8_34. 55

Bernhard Thomaszewski, Stelian Coros, Damien Gauge, Vittorio Megaro, Eitan Grinspun, and Markus Gross. Computational design of linkage-based characters. *ACM Transactions on Graphics*, 33(4):64:1–64:9, July 2014. DOI: 10.1145/2601097.2601143. 31, 52

K. Thrimurthulu, Pulak M. Pandey, and N. Venkata Reddy. Optimum part deposition orientation in fused deposition modeling. *International Journal of Machine Tools and Manufacture*, 44(6):585–594, 2004. DOI: 10.1016/j.ijmachtools.2003.12.004. 10, 63

Skylar Tibbits. 4D printing: Multi-material shape change. *Architectural Design*, 84(1):116–121, 2014. DOI: 10.1002/ad.1710. 90

Pallavi Tiwari, S. Viswanath, J. Kurhanewicz, Akshay Sridhar, and Anant Madabhushi. Multimodal wavelet embedding representation for data combination (maweric): Integrating magnetic resonance imaging and spectroscopy for prostate cancer detection. *NMR in Biomedicine*, 25(4):607–619, 2012. DOI: 10.1002/nbm.1777. 47, 49

Andrew Townsend, N. Senin, Liam Blunt, R. K. Leach, and J. S. Taylor. Surface texture metrology for metal additive manufacturing: A review. *Precision Engineering*, 46:34–47, 2016. DOI: 10.1016/j.precisioneng.2016.06.001. 13, 14

Justin Tyberg and Jan Helge Bøhn. Local adaptive slicing. *Rapid Prototyping Journal*, 4(3): 118–127, 1998. DOI: 10.1108/13552549810222993. 76

Erva Ulu, Emrullah Korkmaz, Kubilay Yay, O. Burak Ozdoganlar, and Levent Burak Kara. Enhancing the structural performance of additively manufactured objects through build orientation optimization. *Journal of Mechanical Design*, 137(11):111410, 2015. DOI: 10.1115/1.4030998. 62

Nobuyuki Umetani and Ryan Schmidt. Cross-sectional structural analysis for 3D printing optimization. In *SIGGRAPH Asia Technical Briefs*, pages 5:1–5:4, 2013. DOI: 10.1145/2542355.2542361. 22, 53, 62

Nobuyuki Umetani and Ryan Schmidt. Surfcuit: Surface-mounted circuits on 3D prints. *IEEE Computer Graphics and Applications*, 38(3):52–60, 2017. DOI: 10.1109/mcg.2017.40. 90

Nobuyuki Umetani, Takeo Igarashi, and Niloy J. Mitra. Guided exploration of physically valid shapes for furniture design. *ACM Transactions on Graphics*, 31(4):86:1–86:11, 2012. DOI: 10.1145/2185520.2335437. 35, 47, 53

Nobuyuki Umetani, Yuki Koyama, Ryan Schmidt, and Takeo Igarashi. Pteromys: Interactive design and optimization of free-formed free-flight model airplanes. *ACM Transactions on Graphics*, 33(4):65:1–65:10, July 2014. DOI: 10.1145/2601097.2601129. 34, 49

Nobuyuki Umetani, Athina Panotopoulou, Ryan Schmidt, and Emily Whiting. Printone: Interactive resonance simulation for free-form print-wind instrument design. *ACM Transactions on Graphics*, 35(6):184:1–184:14, November 2016. http://doi.acm.org/10.1145/2980179.2980250 DOI: 10.1145/2980179.2980250. 35

Mohammad Vaezi and Chee Kai Chua. Effects of layer thickness and binder saturation level parameters on 3D printing process. *International Journal of Advanced Manufacturing Technology*, 53(1-4):275–284, 2011. DOI: 10.1007/s00170-010-2821-1. 85

Juraj Vanek, J. A. Galicia, Bedrich Benes, R Měch, N. Carr, Ondrej Stava, and G. S. Miller. Packmerger: A 3D print volume optimizer. *Computer Graphics Forum*, 33(6), 2014a. DOI: 10.1111/cgf.12353. 47, 50, 53, 60

Juraj Vanek, Jorge A. Garcia Galicia, and Bedrich Benes. Clever support: Efficient support structure generation for digital fabrication. *Computer Graphics Forum*, 33(5):117–125, 2014b. DOI: 10.1111/cgf.12437. 49, 66

Gokul Varadhan and Dinesh Manocha. Accurate Minkowski sum approximation of polyhedral models. *Graphical Models*, 68(4):343–355, 2006. DOI: 10.1016/j.gmod.2005.11.003. 69

Kiril Vidimce, Alexandre Kaspar, Ye Wang, and Wojciech Matusik. Foundry: Hierarchical material design for multi-material fabrication. In *ACM User Interface Software and Technology Symposium*, 2016. DOI: 10.1145/2984511.2984516. 82

Kiril Vidimče, Szu-Po Wang, Jonathan Ragan-Kelley, and Wojciech Matusik. OpenFab: A programmable pipeline for multi-material fabrication. *ACM Transactions on Graphics*, 32(4): 136:1–136:12, 2013. DOI: 10.1145/2461912.2461993. 7, 51, 52, 74, 81

R. Vilar. Laser cladding. *Journal of Laser Applications*, 11:64–79, 1999. DOI: 10.2351/1.521888. 3

Neri Volpato, Alexandre Franzoni, Diogo Carbonera Luvizon, and Julian Martin Schramm. Identifying the directions of a set of 2D contours for additive manufacturing process planning. *International Journal of Advanced Manufacturing Technology*, 68(1-4):33–43, 2013. DOI: 10.1007/s00170-012-4706-y. 83

P. Vuyyuru, C. F. Kirschman, G. Fadel, A. Bagchi, and C. C. Jara-Almonte. A NURBS-based approach for rapid product realization. In *5th International Conference on Rapid Prototyping*, 1992. 79

Pang King Wah, Katta G Murty, Ajay Joneja, and Leung Chi Chiu. Tool path optimization in layered manufacturing. *IIE Transactions*, 34(4):335–347, 2002. DOI: 10.1080/07408170208928874. 84

Charlie C. L. Wang, Yuen-Shan Leung, and Yong Chen. Solid modeling of polyhedral objects by layered depth-normal images on the GPU. *Computer-Aided Design*, 42(6):535–544, 2010. DOI: 10.1016/j.cad.2010.02.001. 78

Charlie C. L. Wang. Approximate boolean operations on large polyhedral solids with partial mesh reconstruction. *IEEE Transactions on Visual Computer Graphics*, 17(6):836–849, 2011. DOI: 10.1109/tvcg.2010.106. 78

Charlie C.L. Wang and Yong Chen. Thickening freeform surfaces for solid fabrication. *Rapid Prototyping Journal*, 19(6):395–406, 2013. DOI: 10.1108/rpj-02-2012-0013. 55

Charlie C. L. Wang and Dinesh Manocha. Efficient boundary extraction of BSP solids based on clipping operations. *IEEE Transactions on Visual Computer Graphics*, 19(1):16–29, 2013a. DOI: 10.1109/tvcg.2012.104. 56

Charlie C.L. Wang and Dinesh Manocha. GPU-based offset surface computation using point samples. *Computer-Aided Design*, 45(2):321–330, 2013b. DOI: 10.1016/j.cad.2012.10.015. 69

Hongyu Wang, Ying He, Xin Li, Xianfeng Gu, and Hong Qin. Polycube splines. *Computer-Aided Design*, 40(6):721–733, 2008. DOI: 10.1016/j.cad.2008.01.012. 51

Lingfeng Wang and Emily Whiting. Buoyancy optimization for computational fabrication. *Computer Graphics Forum*, 35(2):49–58, 2016. DOI: 10.1111/cgf.12810. 14

Weiming Wang, Tuanfeng Y. Wang, Zhouwang Yang, Ligang Liu, Xin Tong, Weihua Tong, Jiansong Deng, Falai Chen, and Xiuping Liu. Cost-effective printing of 3D objects with skin-frame structures. *ACM Transactions on Graphics*, 32(6):177, 2013. DOI: 10.1145/2508363.2508382. 11, 47, 48, 67, 72

Weiming Wang, Haiyuan Chao, Jing Tong, Zhouwang Yang, Xin Tong, Hang Li, Xiuping Liu, and Ligang Liu. Saliency-preserving slicing optimization for effective 3D printing. *Computer Graphics Forum*, 34(6):148–160, 2015. DOI: 10.1111/cgf.12527. 48, 76

Weiming Wang, Cedric Zanni, and Leif Kobbelt. Improved surface quality in 3D printing by optimizing the printing direction. *Computer Graphics Forum*, 35(2):59–70, 2016. DOI: 10.1111/cgf.12811. 12, 61, 76

Weiming Wang, Yong-Jin Liu, Jun Wu, Shengjing Tian, Charlie C. L. Wang, Ligang Liu, and Xiuping Liu. Support-free hollowing. *IEEE Transactions on Visualization and Computer Graphics*, PP(99):1–1, 2017. DOI: 10.1109/tvcg.2017.2764462. 72

Xiangzhi Wei, Siqi Qiu, Lin Zhu, Ruiliang Feng, Yaobin Tian, Juntong Xi, and Youyi Zheng. Toward support-free 3D printing: A skeletal approach for partitioning models. *IEEE Transactions on Visualization and Computer Graphics*, PP(99):1–1, 2017. DOI: 10.1109/tvcg.2017.2767047. 59

Yang Weidong. Optimal path planning in rapid prototyping based on genetic algorithm. In *Chinese Control and Decision Conference*, pages 5068–5072, 2009. DOI: 10.1109/ccdc.2009.5194966. 84

Lee E. Weiss, Robert Merz, Fritz B. Prinz, Gennady Neplotnik, et al. Shape deposition manufacturing of heterogeneous structures. *Journal of Manufacture Systems*, 16(4):239, 1997. DOI: 10.1016/s0278-6125(97)89095-4. 80

H. Wu, E. M. Sachs, N. M. Patrikalakis, D. Brancazio, J. Serdy, and T. R. Jackson. Distributed design and fabrication of parts with local composition control. In *Proc. of the NSF Design and Manufacturing Grantees Conference*, 2000. 82

Jun Wu, Christian Dick, and Rüdiger Westermann. A system for high-resolution topology optimization. *Visualization and Computer Graphics, IEEE Transactions on*, PP(99):1–1, 2015. DOI: 10.1109/tvcg.2015.2502588. 23, 24, 47, 53

Jun Wu, Lou Kramer, and Rüdiger Westermann. Shape interior modeling and mass property optimization using ray-reps. *Computers and Graphics*, 58:66–72, 2016a. DOI: 10.1016/j.cag.2016.05.003. 15

Jun Wu, Charlie C. L. Wang, Xiaoting Zhang, and Rüdiger Westermann. Self-supporting rhombic infill structures for additive manufacturing. *Computer-Aided Design*, 2016b. DOI: 10.1016/j.cad.2016.07.006. 72

Jun Wu, Niels Aage, Rüdiger Westermann, and Ole Sigmund. Infill optimization for additive manufacturing—approaching bone-like porous structures. *IEEE Transactions on Visualization and Computer Graphics*, PP(99):1–1, 2017. DOI: 10.1109/tvcg.2017.2655523. 74

Rundong Wu, Huaishu Peng, François Guimbretière, and Steve Marschner. Printing arbitrary meshes with a 5DOF wireframe printer. *ACM Transactions on Graphics*, 35(4):1–9, 2016c. DOI: 10.1145/2897824.2925966. 80, 81

Y. F. Wu, Y. S. Wong, H. T. Loh, and Y. F. Zhang. Modelling cloud data using an adaptive slicing approach. *Computer-Aided Design*, 36(3):231–240, 2004. DOI: 10.1016/s0010-4485(03)00097-6. 79

Yue Xie and Xiang Chen. Support-free interior carving for 3D printing. *Visual Informatics*, 1 (1):9–15, 2017. DOI: 10.1016/j.visinf.2017.01.002. 72

Shiqing Xin, Chi-Fu Lai, Chi-Wing Fu, Tien-Tsin Wong, Ying He, and Daniel Cohen-Or. Making burr puzzles from 3D models. In *ACM SIGGRAPH Papers*, pages 97:1–97:8, 2011. DOI: 10.1145/1964921.1964992. 38, 52, 53

F. Xu, H. T. Loh, and Y. S. Wong. Considerations and selection of optimal orientation for different rapid prototyping systems. *Rapid Prototyping Journal*, 5(2):54–60, 1999. DOI: 10.1108/13552549910267344. 10

Hongyi Xu, Yijing Li, Yong Chen, and Jernej Barbič. Interactive material design using model reduction. *ACM Transactions on Graphics*, 34(2):18, 2015. DOI: 10.1145/2699648. 25, 47, 50, 53

Wen-Peng Xu, Wei Li, and Li-Gang Liu. Skeleton-sectional structural analysis for 3D printing. *Journal of Computer Science and Technology*, 31(3):439–449, 2016. DOI: 10.1007/s11390-016-1638-2. 14

Ulas Yaman, Nabeel Butt, Elisha Sacks, and Christoph Hoffmann. Slice coherence in a query-based architecture for 3D heterogeneous printing. *Computer-Aided Design*, 75-76:27–38, 2016. DOI: 10.1016/j.cad.2016.02.005. 73

Pinghai Yang and Xiaoping Qian. Adaptive slicing of moving least squares surfaces: Toward direct manufacturing of point set surfaces. *Journal of Computer Information Science in Engineering*, 8(3):031003, 2008. DOI: 10.1115/1.2955481. 79

Pinghai Yang, Kang Li, and Xiaoping Qian. Topologically enhanced slicing of MLS surfaces. In *30th Computers and Information in Engineering Conference, Parts A and B*, Vol. 3, 2010. DOI: 10.1115/detc2010-29125. 79

Y. Yang, H. T. Loh, J. Y. H. Fuh, and Y. G. Wang. Equidistant path generation for improving scanning efficiency in layered manufacturing. *Rapid Prototyping Journal*, 8(1):30–37, 2002. DOI: 10.1108/13552540210413284. 71

Yang Yang, Shuangming Chai, and Xiao-Ming Fu. Computing interior support-free structure via hollow-to-fill construction. *Computers and Graphics*, 2017. DOI: 10.1016/j.cag.2017.07.005. 72

Jiaxian Yao, Danny M. Kaufman, Yotam Gingold, and Maneesh Agrawala. Interactive design and stability analysis of decorative joinery for furniture. *ACM Transactions on Graphics*, 36(2):20:1–20:16, March 2017. http://doi.acm.org/10.1145/3054740 DOI: 10.1145/3072959.3126857. 39

Miaojun Yao, Zhili Chen, Linjie Luo, Rui Wang, and Huamin Wang. Level-set-based partitioning and packing optimization of a printable model. *ACM Transactions on Graphics*, 34(6): 214:1–214:11, 2015. DOI: 10.1145/2816795.2818064. 47, 50, 53, 60

Christopher Yu, Keenan Crane, and Stelian Coros. Computational design of telescoping structures. *ACM Transactions on Graphics*, 36(4):83:1–83:9, July 2017. DOI: 10.1145/3072959.3073673. 30

Kemal Yuksek, Weihan Zhang, Boryslaw Iwo Ridzalski, and Ming C. Leu. A new contour reconstruction approach from dexel data in virtual sculpting. In *3rd International Conference on Geometric Modeling and Imaging*, 2008. DOI: 10.1109/gmai.2008.27. 78

Jelena Žarko, Gojko Vladić, Magdolna Pál, and Sandra Dedijer. Influence of printing speed on production of embossing tools using FDM 3D printing technology. *Journal of Graphic Engineering and Design*, 8(1):19, 2017. DOI: 10.24867/jged-2017-1-019. 84

ZCorporation. Zprinter 450 hardware manual. http://www.3dsystems.com/, 2007. 85

Jonas Zehnder, Stelian Coros, and Bernhard Thomaszewski. Designing structurally-sound ornamental curve networks. *ACM Transactions on Graphics*, 35(4):99:1–99:10, July 2016. DOI: 10.1145/2897824.2925888. 20, 48, 50, 51, 53

Long Zeng, Lip Man-Lip Lai, Di Qi, Yuen-Hoo Lai, and Matthew Ming-Fai Yuen. Efficient slicing procedure based on adaptive layer depth normal image. *Computer-Aided Design*, 43 (12):1577–1586, 2011. DOI: 10.1016/j.cad.2011.06.007. 78

Hao Zhang, Oliver Van Kaick, and Ramsay Dyer. Spectral mesh processing. *Computer Graphics Forum*, 29(6):1865–1894, 2010. DOI: 10.1111/j.1467-8659.2010.01655.x. 49

Ran Zhang, Thomas Auzinger, Duygu Ceylan, Wilmot Li, and Bernd Bickel. Functionality-aware retargeting of mechanisms to 3D shapes. *ACM Transactions on Graphics*, 36(4):81:1–81:13, July 2017. DOI: 10.1145/3072959.3073710. 31

Weihan Zhang and Ming C. Leu. Surface reconstruction using dexel data from three sets of orthogonal rays. *Journal of Computer Information Science in Engineering*, 9(1):011008, 2009. DOI: 10.1115/1.3086034. 78

Weihan Zhang, Xiaobo Peng, Ming C. Leu, and Wei Zhang. A novel contour generation algorithm for surface reconstruction from dexel data. *Journal of Computer Information Science in Engineering*, 7(3):203–210, 2007. DOI: 10.1115/1.2752817. 78

Xiaolong Zhang, Yang Xia, Jiaye Wang, Zhouwang Yang, Changhe Tu, and Wenping Wang. Medial axis tree—an internal supporting structure for 3D printing. *Computer Aided Geometric Design*, 35-36:149–162, 2015a. DOI: 10.1016/j.cagd.2015.03.012. 72

Xiaoting Zhang, Xinyi Le, Athina Panotopoulou, Emily Whiting, and Charlie C. L. Wang. Perceptual models of preference in 3D printing direction. *ACM Transactions on Graphics*, 34 (6):215:1–215:12, 2015b. DOI: 10.1145/2816795.2818121. 13, 61, 68

Yicha Zhang, Alain Bernard, Ramy Harik, and K. P. Karunakaran. Build orientation optimization for multi-part production in additive manufacturing. *Journal of Intelligent Manufacturing*, pages 1–15, 2015c. DOI: 10.1007/s10845-015-1057-1. 60

Yizhong Zhang, Chunji Yin, Changxi Zheng, and Kun Zhou. Computational hydrographic printing. *ACM Transactions on Graphics*, 34(4):131:1–131:11, July 2015d. DOI: 10.1145/2766932. 17

Zhengyan Zhang and Sanjay Joshi. An improved slicing algorithm with efficient contour construction using STL files. *International Journal of Advanced Manufacturing Technology*, 80 (5-8):1347–1362, 2015. DOI: 10.1007/s00170-015-7071-9. 77

Haiming Zhao, Weiwei Xu, Kun Zhou, Yin Yang, Xiaogang Jin, and Hongzhi Wu. Stress-constrained thickness optimization for shell object fabrication. *Computer Graphics Forum*, 36 (6):368–380, 2017. DOI: 10.1111/cgf.12986. 56

Haisen Zhao, Baoquan Chen, Fanglin Gu, Qi-Xing Huang, Jorge Garcia, Yong Chen, Changhe Tu, Bedrich Benes, Hao Zhang, and Daniel Cohen-Or. Connected Fermat spirals for layered fabrication. *ACM Transactions on Graphics*, 35(4):1–10, 2016a. DOI: 10.1145/2897824.2925958. 48, 50, 71, 83, 84

Haisen Zhao, Lin Lu, Yuan Wei, Dani Lischinski, Andrei Sharf, Daniel Cohen-Or, and Bao-quan Chen. Printed perforated lampshades for continuous projective images. *ACM Transactions on Graphics*, 35(5):154:1–154:11, June 2016b. DOI: 10.1145/2907049. 17

Hanli Zhao, Charlie C. L. Wang, Yong Chen, and Xiaogang Jin. Parallel and efficient boolean on polygonal solids. *Visual Computer*, 27(6-8):507–517, 2011. DOI: 10.1007/s00371-011-0571-1. 78

Chi Zhou, Yong Chen, Z. G. Yang, and Behrokh Khoshnevis. Development of multi-material mask-image-projection-based stereolithography for the fabrication of digital materials. In *Annual Solid Freeform Fabrication Symposium*, 2011. 4

Qingnan Zhou, Julian Panetta, and Denis Zorin. Worst-case structural analysis. *ACM Transactions on Graphics*, 32(4):137:1–137:12, 2013. DOI: 10.1145/2461912.2461967. 14, 15, 22, 47, 50, 53

Shiwei Zhou and Qing Li. Microstructural design of connective base cells for functionally graded materials. *Materials Letters*, 62(24):4022–4024, 2008. DOI: 10.1016/j.matlet.2008.05.058. 74

Shizhe Zhou, Changyun Jiang, and Sylvain Lefebvre. Topology-constrained synthesis of vector patterns. *ACM Transactions on Graphics*, 33(6):215:1–215:11, November 2014a. DOI: 10.1145/2661229.2661238. 19

Yahan Zhou, Shinjiro Sueda, Wojciech Matusik, and Ariel Shamir. Boxelization: Folding 3D objects into boxes. *ACM Transactions on Graphics*, 33(4):71:1–71:8, 2014b. DOI: 10.1145/2601097.2601173. 36, 47, 52, 60

Bo Zhu, Mélina Skouras, Desai Chen, and Wojciech Matusik. Two-scale topology optimization with microstructures. *ACM Transactions on Graphics*, 36(5):164:1–164:16, July 2017. DOI: 10.1145/3072959.3126835. 27, 74

Lifeng Zhu, Weiwei Xu, John Snyder, Yang Liu, Guoping Wang, and Baining Guo. Motion-guided mechanical toy modeling. *ACM Transactions on Graphics*, 31(6):127:1–127:10, November 2012. DOI: 10.1145/2366145.2366146. 30, 52

W. M. Zhu and K. M. Yu. Dexel-based direct slicing of multi-material assemblies. *International Journal of Advanced Manufacturing Technology*, 18(4):285–302, 2001. DOI: 10.1007/s001700170069. 78, 80

Qiang Zou, Juyong Zhang, Bailin Deng, and Jibin Zhao. Iso-level tool path planning for free-form surfaces. *Computer-Aided Design*, 53:117–125, 2014. DOI: 10.1016/j.cad.2014.04.006. 71, 83

Authors' Biographies

MARCO ATTENE

Marco Attene is a permanent researcher at CNR-IMATI in Genova, where he studies geometry processing for 3D printing applications. He has a Ph.D., a research management diploma, and a full professorship habilitation. Marco has been the principal investigator at CNR for regional, national, and international projects, has or had collaborations and joint research programs with both industry and academy in Europe, the U.S., Asia, and New Zealand, and serves as an evaluator for research funding agencies. Marco has published high-impact articles in prestigious journals in the area, and contributed to the organization of international conferences as program chair, program committee member, and organizing committee member. He is an associate editor of international journals in the area and is an active software developer (his "MeshFix" system received the SGP Software Award in 2014). Marco is currently leading the *Process Planning* work package in the EU H2020 project *CAxMan* on Additive Manufacturing.

MARCO LIVESU

Marco Livesu is a permanent researcher at the Institute for Applied Mathematics and Information Technologies of the National Research Council of Italy (CNR IMATI). He received his Ph.D. at the University of Cagliari (2010–2014), after which he was a postdoctoral researcher at the University of British Columbia (2014–2015), University of Cagliari (2015), and CNR IMATI (2015–2016). His main research interests are digital modeling for manufacturing and mesh generation. In 2015, he was one of the recipients of the ERCIM Alain Bensoussan PostDoctoral Fellowship. He is the creator and developer of CinoLib, a C++ library for processing general polygonal and polyhedral meshes. Marco is currently involved in the EU H2020 project *CAxMan* on Additive Manufacturing.

SYLVAIN LEFEBVRE

Sylvain Lefebvre is a senior researcher at Inria (France). After a Ph.D. at Grenoble Alpes University (2002–2004), he joined Microsoft Research as a postdoctoral researcher in 2005 and Inria in 2006. His main research focus is to simplify content creation, synthesizing highly detailed patterns, structures and shapes, with applications in computer graphics and additive manufacturing. Sylvain received the EUROGRAPHICS Young Researcher Award in 2010 for his work on texturing data structures and runtime procedural texture synthesis. Since 2012, he has been the principal investigator of the ERC Shape-Forge project, which focuses on shape synthesis for additive manufacturing. He created and is the lead developer of the IceSL software for modeling for additive manufacturing. Sylvain is an active member of the community and serves on the technical papers committee and editorial boards of the main conferences and journals of the field.

THOMAS FUNKHOUSER

Thomas Funkhouser is a Professor in the Department of Computer Science at Princeton University. He received a B.S. in Biological Sciences from Stanford University in 1983, a M.S. in Computer Science from UCLA in 1989, and a Ph.D. in Computer Science from UC Berkeley in 1993. He has published more than 100 research papers and received several awards, including the ACM SIGGRAPH Computer Graphics Achievement Award. His research focuses on 3D shape analysis, geometric modeling, and scene understanding.

SZYMON RUSINKIEWICZ

Szymon Rusinkiewicz is a Professor of Computer Science at Princeton University. His work focuses on the interface between computers and the visual and tangible world: acquisition, representation, analysis, and fabrication of 3D shape, motion, surface appearance, and scattering. He investigates algorithms for processing geometry and reflectance, including registration, matching, completion, hierarchical decomposition, symmetry analysis, sampling, and depiction. Applications of this work include documentation of cultural heritage artifacts and sites, appearance and performance capture for digital humans, and illustrative depiction through line drawings and non-photorealistic shading models.

STEFANO ELLERO

Stefano Ellero is Project Manager at Stam S.r.l. He is a well-experienced mechanical engineer with expertise in design, engineering, prototyping and testing of mechanical devices, mechatronic systems, and production processes. He has strongly contributed to several developments of Stam in applications such as: automotive, bio-engineering, robotics, and aerospace. He is one of the inventors of the NUGEAR gearbox, use-case demonstrator of the CAxMan project, where Stam is partner and Stefano is actively involved.

JONÀS MARTÍNEZ

Jonàs Martínez is a researcher at Inria. He received a Ph.D. from Universitat Politècnica de Catalunya in 2013, and was awarded an ERCIM fellowship (FP7 Marie Curie actions). Since then he worked in the frame of the ERC Shapeforge project, which aims at helping users design new, complex objects from examples, in the context of additive manufacturing. His current recent research lies at the intersection between fabrication, computer graphics, and geometry processing. He has published several papers on these topics at high-impact journals and conferences and contributed algorithms to the IceSL modeler and slicer for additive manufacturing.

AMIT HAIM BERMANO

As of 2017, Dr. **Amit H. Bermano** is a postdoctoral researcher at the Princeton Graphics Group, Princeton University. Previously, he was a postdoctoral researcher at Disney Research Zurich in the computational materials group (2016). He conducted his doctoral studies at ETH Zurich, in collaboration with Disney Research Zurich (2016). His Masters and Bachelors degrees were obtained at The Technion–Israel Institute of Technology. His research focuses on connecting the geometry processing field with other fields in computer graphics and vision, mainly by using geometric methods to facilitate other applications. His interests in this context include computational fabrication, animation, augmented reality, medical imaging, and machine learning.

Printed in the United States
by Baker & Taylor Publisher Services